DC AMPLIFIERS IN
INSTRUMENTATION

DC AMPLIFIERS IN INSTRUMENTATION

RALPH MORRISON

Astrodata, Inc.
Anaheim, California

WILEY-INTERSCIENCE

A Division of John Wiley & Sons, Inc.
New York · London · Sydney · Toronto

Copyright © 1970, by John Wiley & Sons, Inc.

10 9 8 7 6 5 4 3 2 1

Library of Congress Catalogue Card Number: 73-116766

ISBN 0-471-61600-1

Printed in the United States of America

Dedicated
to
my wife Lee

Preface

The dc amplifier as an electronic building block has developed rather rapidly in the last 20 years. To the user of instrumentation this growth has led to a lot of confusion. To the uninitiated the problems seem trivial and overcomplicated; but the reason for the confusion is partly related to the closed nature of the instrumentation business. Much of the evolution, except in the field of *IC* technology, has been accomplished by individuals in small aggressive instrumentation houses. Out of necessity these houses have considered much of their technology proprietary, and as a result no consistent method has existed for keeping the engineering community fully abreast of developments. The new engineer has been left to his own devices if and when information happens to come along.

Dc amplifier evolution has been strongly related to component technology. As new components evolve, designers have been quick to apply them to reduce cost and improve performance specifications. Not all innovation, however, has been component-oriented. Many philosophical changes have taken place over the years. Today's ideas could have worked with yesterday's components, but the way was just not that clear.

Dc amplifiers change to meet market demand. Currently, instruments are being designed for remote computer control and with automatic gain ranging. Tomorrow the techniques may involve a complete system integration. Whatever the path taken in design, specifications, dynamic response, and performance evaluation will remain nearly constant.

I have spent some 20 years in the "care and feeding" of dc amplifiers. This effort was made in two small businesses (a dollar volume under two million dollars). Early experiences included vacuum-tube designs, for which one amplifier and its associated power supply could *not* be carried by one man. Today a complete unit can easily be carried in the palm of one hand. Early design contributions included a multiloop feedback circuit to eliminate the need for highly regulated power supplies.

Later a post high-frequency modulator scheme was developed to reject high-level common-mode voltages. Each advance in design philosophy required a new customer contact effort through sales and instruction material. This combination of design effort, customer liaison, and writing was the background for writing this book.

The practical world of feedback and component selection is not available to everyone. By necessity the designer gets close to these problems. The world of half-defined or ill-defined engineering terms makes specification reading and writing complex. The people closest to the design know how to "play the game." I have not encountered every component idiosyncrasy nor have I seen every specification problem. The hope is that the material in this book will help the engineer, the technician, and the user to "play the game." The engineer should find the material useful in new design efforts. The user should find it an aid to understanding why and how one design differs from another. The book should qualify the reader to interpret specifications and to evaluate performance. Hopefully, the reader may be better equipped to measure what has been intentionally left unsaid in a specification.

The book starts with a discussion of simple components. Semiconductors are well covered in the literature, and for this reason no effort has been made to discuss their characteristics in any detail. Resistors and capacitors are discussed because they are used as feedback elements and their characteristics limit instrument performance. Transformers and inductors are treated in detail for two reasons. First, the shielding of transformers is the key to a good instrument, and, second, the designer is often required to specify or hand-wire first items to prove feasibility.

Feedback theory is presented mostly on a nonmathematical basis. The benefits and limitations of feedback are discussed in detail, and potentiometric and operational feedback circuits are outlined with practical gain and offset techniques.

Many low-frequency, low-level dc amplifiers use modulation and ac amplification. The details of carrier-type processes are presented with various feedback techniques.

Instrumentation amplifiers are discussed mainly in block form so that the shielding processes are clear. Specification limitations for each type of design are covered, and common-mode signals and their rejection are considered in detail.

Specifications and their interpretation are covered in the last chapters. Many subtle points in measurement are brought out to illustrate the difficulties encountered by the engineer and the pitfalls hidden in the specifications awaiting the user.

Many thanks go to many people for making this book possible. An understanding wife accepted the long hours of writing without complaint. Grace Eckmark provided a very helpful hand in doing the typing. A thank you goes to the publisher for his confidence in publishing this effort.

RALPH MORRISON

Pasadena, California
April 1970

Contents

DC AMPLIFIERS IN
INSTRUMENTATION

1

Linear Components

1.1 RESISTORS

Resistors are imperfect devices. The degree and nature of their imperfections vary with the style and type of resistor considered. Since resistors are usually the limiting factor in instrument design, it is important to be very familiar with their performance characteristics. Compensating procedures can often be applied to side-step difficulties. The degree of compensation and the performance required both enter into the selection of a best resistor type.

1.2 CARBON RESISTORS

Molded-carbon resistors are used extensively in discrete-component circuitry. Standard wattages vary from $\frac{1}{8}$ to 2 W, ohmic values from 4.3 Ω to 100 MΩ, and tolerances vary from 5 to 20%. As plentiful as carbon resistors are, engineers are often unfamiliar with their limitations.

1.3 NOISE

The thermal agitation noise of carbon resistors is very close to the theoretical limit given by the equation

$$E_N{}^2 = 4kT \, \Delta f R \tag{1}$$

where E_N is the rms noise, T is the absolute temperature in degrees Kelvin, Δf is the frequency band of interest, and k is Boltzmann's constant equal to 1.38×10^{-23} J/°K.

The noise in carbon resistors increases when the resistors are stressed mechanically or electrically. Low current densities or moderate voltage levels do not add significantly to the thermal noise. It is correct to use carbon resistors as circuit elements in the input stages of most amplifiers. The nonlinearities encountered in the active elements of the circuit usually far exceed the nonlinearities found in carbon resistors. Carbon resistors should not generally be used as feedback elements.

1.4   VOLTAGE COEFFICIENT OF RESISTANCE

Carbon resistors change ohmic value as a function of electrical stress. The effect is usually less than $\frac{1}{4}\%$ and varies as a function of wattage, resistance value, and voltage. This voltage coefficient of resistance is usually negative; that is, the resistance tends to fall as the voltage is increased. A typical normalized resistance curve is shown in Figure 1.1. The curve assumes no resistance change caused by heating.

1.5   REPEATABILITY

Carbon resistors generally stay within a $\pm 5\%$ value range for extended periods of time. If they are operated at elevated temperatures or above their rated wattages, they do not always return to the same resistance value. The degree of variation is statistical and a function of resistance.

1.6   GENERAL APPLICATION NOTES

Many designers hold the actual dissipation maximum to below one-half of the rated wattage. This practice is safe and ensures reasonable

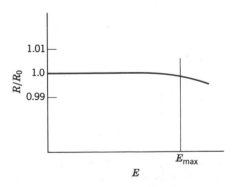

Figure 1.1   A typical voltage coefficient of resistance for a $\frac{1}{2}$-W carbon resistor.

freedom from hot spots. The temperature rise in a given situation is more significant than the wattage ratings. Resistors can withstand tremendous peak-power loads as long as they do not overheat internally. If the packing density is high, the key problem may be heat removal rather than sufficient resistor wattage. Wattage implies heat, and heat produces a temperature rise. Damage eventually results if the heat-removal rate is inadequate.

Carbon resistors are generally constructed with a molded shell around the resistive element. The molded shell mechanically supports the resistor leads. If the lead is stressed sufficiently, a noisy lead-to-element connection may result. For this reason leads should never be cut or bent with the operating tool in contact with the resistor body. The forces that are available can exert damaging stress on the internal connections of the resistor. These same forces can result if a vertically mounted resistor has its body in direct contact with a circuit board. If, after soldering, the body of the resistor is pushed to one side, the forces on the lead can rip a solder joint apart or disengage the lead from the carbon element. To aid in assembly, small soluble plastic rings are available for spacing component bodies away from circuit boards. A simple cleaning rinse dissolves the plastic spacers and the risk of mechanical damage is removed.

Carbon resistors have very low series inductance, and for this reason they make excellent high-frequency elements. They have, however, a shunt capacitance that varies with wattage. This capacitance is not a simple element and must be considered as a distributed parameter. It is obvious that lead dress and lead length effect the value of the shunt capacitance. In a $\frac{1}{4}$-W resistor, the capacitance is between 1 and 2 pF.

1.7   WIREWOUND RESISTORS

Many types and sizes of wirewound resistors are available on the market. Manufacturers will often specialize in certain specification areas. If price and just generally good performance are the considerations, the choice of manufacturer expands considerably. There are subtleties in the manufacturing processes, which account for many of the differences between sources.

Most wirewound resistor manufacturers keep a large stock of resistance wire on hand. They tabulate the sensitivity and also the temperature coefficients of each spool of wire, and they wind their most precise and stable resistors from this stock. Because stress applied to the wire during winding will affect the terminal resistance, most precision resistors are wound under controlled low tension. The resistors are then temperature cycled to remove any accumulated stress. For very accurate and stable resistor values, trimming occurs after cycling. In some plants the resistance change caused by temperature cycling can be predicted, and the resistors are trimmed accordingly when first wound. If resistors are not temperature cycled in manufacture, they can change value under the conditions of normal usage.

The temperature coefficient of resistance (TC) varies slightly between various spools of wire. It is possible to match resistors for TC by winding them from the same spool or by matching spools if the wire size differs. To match or reduce a TC, it is possible to wind resistors from two different spools. The use of these techniques obviously adds to the price of the resistors.

1.8   PARASITIC EFFECTS IN WIREWOUND RESISTORS

A wirewound resistor represents a complex network to ac signals. The lowest frequency at which the network must be considered is dependent on the accuracy requirements and on the resistance value involved. In general the coiling of wire introduces inductance and the coil-to-coil proximity introduces capacitance. The actual network is distributed in nature and can be approximated as a lumped parameter network for evaluation.

The inductive effects can be reduced by controlled reversals of winding direction. If one reverse is used, the reduction in inductance depends on the mutual inductance between the two coils.

$$L_T = L_1 + L_2 - 2L_{12} \qquad\qquad\qquad\qquad (2)$$

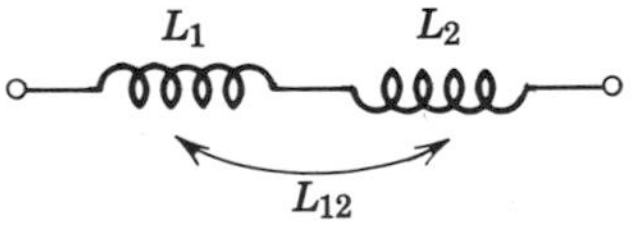

Figure 1.2

The coupling coefficient k is defined as

$$k = \frac{L_{12}}{(L_1 - L_2)^{\frac{1}{2}}} \qquad (3)$$

For 100% coupling, $k = 1$. If $L_1 = L_2$ and $k = 1$, then $2L_{12} = 2L_1$ and $L_T = 0$. This ideal can be only approximated in practice. The use of more than one reversal in turn winding direction generally increases the value of k.

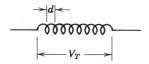

Figure 1.3 A typical coil of resistance wire.

The capacitances involved in a wirewound resistor are distributed in nature. The capacitance is very dependent on the geometry and the method used in winding the coils. This becomes clear if the electrostatic energy stored in the electric field is evaluated.

Consider a coil that is linearly spaced along a bobbin with n turns and a voltage V_T across the coil as in Figure 1.3. The capacitance between any two turns is approximately proportional to the reciprocal of the spacing d between the turns. If C is the capacitance between turns separated by a unit distance, then the energy E_d stored between any two turns is given by

$$E_d = \tfrac{1}{2}\frac{CV_d{}^2}{d} \tag{4}$$

where V_d is the voltage between turns and d is the spacing. The total energy stored in the field is the sum of energies stored between every pair of turns. Once the total energy storage E_T is known, the effective capacitance C_E is related by energy equivalence or

$$E_T = \tfrac{1}{2}C_E V_T{}^2 = \sum_{\substack{\text{all}\\ \text{turns}}} E_d = \sum_{\substack{\text{all}\\ \text{turns}}} \tfrac{1}{2}\frac{CV_d{}^2}{d} \tag{5}$$

The energy summation (5) is proportional to the square of the applied voltage V_t. Solving for C_T in (5) and noting that the square of voltage appears on both sides of the equation, we obtain

$$C_E = \frac{k}{V_t{}^2}\sum E_d \tag{6}$$

To keep C_E low, the energy storage E_d in (4) must be small. This implies that turn-pairs with the largest potential difference should be kept furthest apart. The optimum construction is to have the coil proceed uniformly across the resistor. This is difficult in practice, and the compromise that is often used spaces several segments (pi sections) along the resistor length. Adjacent segments might be wound in opposite directions to keep the inductive effects minimum.

1.9   EQUIVALENT CIRCUITS

The parasitic effects on a wirewound resistor can be approximated by the network shown in Figure 1.4 where there are n sections to the resistor. The impedance of this network is

$$Z = \frac{n(Ls + R)}{LCs^2 + RCs + 1} \tag{7}$$

where s is the complex-frequency variable. Typical values might be $R = 10 \text{ k}\Omega$, $L = 0.5 \text{ mH}$, and $C = 1.0 \text{ pF}$.

The resonant properties of the network dominate if R is low in value. The impedance Z will have a maximum near the resonant frequency. If R is very large, the denominator of Z is dominated by the middle term. If $L/C = R^2$, the impedance can be held essentially constant over a wide frequency range.

Wirewound precision resistors are usually made to order with the user specifying the value, accuracy, frequency response, wattage, and size. Unfortunately these specifications are not mutually exclusive: for example, small size and high wattage are not compatible; high accuracy and stability usually require the use of heavier wire and these elements are therefore larger.

High-frequency resistors are a specialty for some manufacturers. These resistors are usually longer in length to reduce the capacitance and inductance. The term "high frequency" is a relative term, as the following examples illustrate. A 10-MΩ wirewound resistor manufactured with careful attention to high-frequency characteristics can have a 0.1% error at 100 Hz. A 1000-Ω high-frequency resistor used as a feedback element in a 10-mHz amplifier can cause the amplifier to oscillate (a carbon resistor used in the same place provides stable performance).[1] It should be apparent that wirewound resistors can be difficult to apply to high-fre-

---

[1] A carbon resistor may be inadequate for other reasons. Film resistors for this application are discussed in Section 1.12.

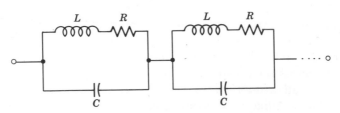

Figure 1.4   Equivalent network of a wirewound resistor.

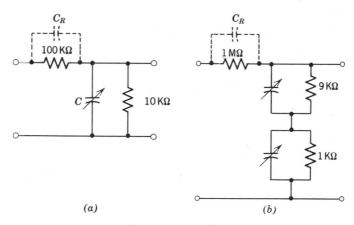

Figure 1.5   Attenuator compensation.

quency problems. Two resistors in series form an attenuator. Attenuator compensation can be made over limited ranges of frequency. First-order attenuator compensation involves a shunting capacitor across the smaller resistor. Second-order compensation might include a series of RC tank circuits where R is less than 1% of the resistor being compensated. This type of compensation for a parasitic capacitance C_R is shown in Figure 1.5. To test an attenuator for balance, square waves should be used. When the response is exactly flat, the attenuator is compensated.

1.10   FOUR-TERMINAL RESISTORS

In many applications the exact point on the leads which corresponds to the specified resistance is critical. If additional connections are brought out at these exact points, the resistor is said to be "four-terminaled." This treatment is usually reserved for low-value resistors as here the leads are a significant percentage of the resistance value.

The leads that are brought out at the point of exact value also have a resistance, but as long as they do not carry a large current the potential drop V measured on them will equal IR, where R is the exact value of the resistor. Figure 1.6a illustrates this practice with terminals 2 and 3 monitored only for voltage. Note that the terminal pairs are not interchangeable. R_s is the external undesirable lead resistance. Accuracy dictates whether the four-terminal approach is necessary; for example, a 0.1% 10-Ω resistor is best treated in a four-terminal manner because 0.1% corresponds to 10 mΩ of lead resistance, a value easily picked

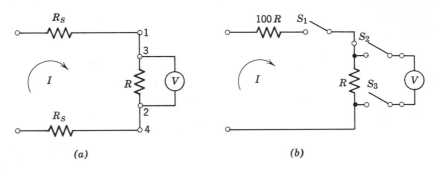

Figure 1.6 A four-terminal resistor.

up in wiring. It follows that 0.01% and 100 Ω should also be treated in a four-terminal manner as the error resistance here is 10 mΩ.

Precautions should be taken when using low-valued resistors. Their location, wire routing, methods of switching, etc., all become important. Low-impedance circuits can permit large circulating currents from coupled magnetic fields. These circuits should therefore be located away from transformers, blowers, and relays. If distances are significant, wiring resistance can add enough resistance to place the total resistor out of tolerance. It is possible to switch low-valued resistors as four-terminaled components. Switch contacts can be positioned so that they do not affect the accuracy requirement, as shown in Figure 1.6*b*.

1.11   HIGH RESISTANCE VALUES

The largest value of standard carbon resistor is 100 MΩ. Wirewound resistors up to 10 MΩ resistance can be manufactured, but at these values the resistors become both bulky and costly. For resistance values higher than 100 MΩ, film techniques are usually used. Some manufacturers deposit a thin film of carbon on glass and then cut the deposit in a spiral shape to increase the path length and decrease the cross-sectional area. Often the deposited film is held in an inert environment to protect the surface against contamination.

Accuracy becomes a serious problem for high resistance values. In the range 10^{12} Ω 1% accuracy is attainable. For 10^{13} Ω an accuracy better than 2% is difficult. For resistors greater than 10^{13} Ω the methods of measurement become quite indistinct and the user may have to devise his own techniques.

The difficulty becomes immediately apparent if one recalls a few facts. For 10^{14} Ω a 1-pF shunt capacitance corresponds to an RC time constant of 100 sec. The assumption that the capacitance is simple and parallels the resistor body is not valid. In general, the capacitance is distributed in nature and end effects must be considered.

A second difficulty involves current levels. For 100 V across 10^{14} Ω the dc flow is only 1 pA. A measure of current to determine resistance might be invalid because it does not consider the voltage coefficient or temperature coefficient of resistance. The manufacturer may have selected a different value of voltage and temperature for his test and the two results will not agree.

The typical voltage coefficient for resistors up to 10^{10} Ω is $-0.02\%/$V. A 100-V potential drop causes a resistance change downward of 2%. For resistors above 10^{10} Ω the voltage coefficient usually gets much worse; for example, a 10^{13}-Ω resistor may have a voltage coefficient of $-0.12\%/$V. Typical temperature coefficients are $0.05\%/$°C at 10^{8} Ω, $0.08\%/$°C at 10^{10} Ω, and $0.23\%/$°C at 10^{13} Ω.

The parasitic capacitance effects on high-value resistors make them difficult to handle at high frequencies. A 100-MΩ resistor can be compensated to provide 10% operation up to about 50 kHz. Typically the shunting effect of 1 pF causes the impedance to drop 3 dB at 10 kHz. Beyond 10 kHz the distorted effects become dominant and compensation becomes very difficult.

High-value resistors must be mounted so that they are removed or shielded from other elements or circuitry. Insulating materials that touch the resistor body can affect dynamic behavior. Circuits near the resistor can easily reactively couple to the resistor. Placing the resistor on an insulating board will increase the shunt capacitance because of the higher dielectric constant of the board over air. I once found that a particular gummed label placed on a glass resistor body changed the dynamic behavior of the resistor decidedly. With the label removed, everything was fine.

High-value resistors are easily shunted in value by contamination gathering on the resistor body or on the mounting surface. The precautions to be taken in design include proper covering or potting as required. Potting is ineffective unless at least one terminal of the resistor can be potted with the resistor. If this is not done, the potting surface merely serves to provide another (longer) shunting path around the resistor.

1.12 FILM RESISTORS

Resistors having a film of metal as a resistance element are usually called metal-film resistors. A thin film of several metals is varied in composition to accommodate wattage, resistance values, and TC. The path length and cross section are determined by a spiral path cut into the deposited element. Resistors cover the range 10 Ω to 30 MΩ but not in all wattage and tolerance sizes.

The voltage coefficient of resistance for metal-film resistors below 10 kΩ is very small. For resistors above 10 kΩ, the coefficient rarely exceeds 5 ppm/V. This fact makes the metal-film resistor more attractive than all carbon resistors. The temperature coefficient of resistance of metal-film resistors can be specified to fall within certain limits over wide temperature ranges; for example, the tolerance ±25 ppm can be held over the temperature span −55 to 165°C and is available for the total range of resistance values ordered. This matching of TC is a desirable feature in many circuits such as feedback elements and attenuators.

Metal-film resistors can be ordered with tight tolerances; for example, some wattages can be supplied with a tolerance of ±0.05% although $\frac{1}{2}$, 1, and 2% values are more common. The accuracies available are an indication of the stability of these metal-film elements. Resistors can be made to order set to a special value. Many manufacturers cover the range in 2% values. This spacing accommodates most requirements without the need for special values.

Metal-film resistors are usually superior to wirewound resistors for frequency response and rise time, particularly for the larger resistance values. The coil inductance problem of wirewound resistors is not a problem for metal-film resistors.

Early film resistors were made of a deposited carbon. Such a resistor is relatively inexpensive and is entirely adequate for many applications. The principal differences between it and a metal-film resistor are that for carbon-film resistors:

1. Resistor values are available up to 100 MΩ.

2. The TC is negative and varies significantly between resistors of the same wattage and resistance value.

3. The TC for low-value resistors varies between −150 and −350 ppm/°C. Above 1 MΩ the TC can be as high as −3500 ppm/°C.

Thick-film deposition can be used for resistors operating at high power. These resistors usually have specifications relating to the cycling of power and temperature that are superior to their equivalent in metal-film or carbon-film components.

1.13   POTENTIOMETERS—GENERAL DISCUSSION

Potentiometers come in a variety of shapes and sizes to fit numerous applications. Since it is not practical to discuss the pros and cons of specific styles or brands, this section deals with the most frequently encountered troubles.

A potentiometer is a resistor with a slider added. The resistance element can be wire, carbon, metal film, carbon film, etc. All of the discussion concerning resistors applies. Inductance is not removed simply from a wirewound potentiometer as the geometry requires a lower coupling coefficient (see discussion in Section 1.8). The potentiometer manufacturer has two additional problems to face and these are resolution and slider wear and tear; for example, five turns of resistance wire is acceptable in a resistor, but in a potentiometer the number of turns usually exceeds 1000, and the wire must be heavy enough to withstand the abuse of the sliding contact. Obviously the size of the potentiometer must be increased if ruggedness and reliability are key factors. Multiturn potentiometers use a spiral coil of wire and they can accommodate large numbers of turns of reasonably sized wire.

1.14 CAPACITANCE EFFECTS

The outer cover of a potentiometer can be metal or plastic. Metal-cased units are usually mounted with the case in contact with the shield structure of the circuit. The shielding adds a parasitic distributed capacitance from the element to signal common. Here circuit performance can vary as a function of slider position; for example, the variable gain control of a wide-band amplifier can cause variations in frequency response as the gain control is adjusted.

Breadboards often omit the function of controls as an expediency measure, and problems do not appear until the final component or geometry is present. The important lesson to be learned here is that the design prototype should treat the potentiometer problem as close to final form as is practical. This is a matter of judgment, as not all applications are equally critical.

The ungrounded case or plastic-case application may require external shielding as the element's coupling to nearby circuitry may be troublesome. Here again specifics dictate the form of treatment. An uncased unit may be necessary or desirable as the parasitic capacitances are considerably lower in such a unit.

Capacitances can sometimes be partially compensated for by an added capacitor from the slider to one side of the potentiometer. This capacitor will have its maximum effect at the center resistance position and will be completely ineffective when shorted out at one end position of the slider (see Figure 1.7 for an equivalent circuit).

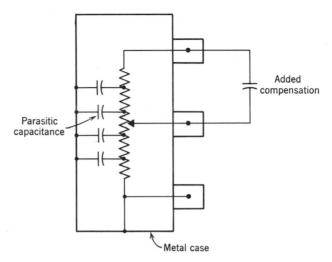

Figure 1.7 Capacitance associated with a potentiometer.

1.15   TAPER

In many applications, resolution requirements vary as a function of angular setting. Several types of potentiometers are available with tapered resistance elements to meet this need. Wirewound potentiometers with nonlinear tapers are available but at considerable cost. A compromise that is sometimes used involves winding segments of the potentiometer with varying sizes of wire with the taper unchanged over each segment. Only a few standard arrangements are available, usually the result of a specific application requirement. In tapered wirewound controls the resolution over each segment can vary significantly, and this variation should be carefully considered.

Standard tapers available in some carbon controls are shown in Figure 1.8. An example in which log taper is required is in volume controls as the ear hears in a logarithmic manner. Many controls require a reciprocal relationship; for example, gain in a feedback amplifier might be given by the expression $G = K/R_1$ and it is desirable to vary R_1 such that G varies linearly with control rotation θ. A control with nearly reciprocal taper implies that $R \simeq K/\theta$ over a specific range of θ, which exactly fits the requirement. In some circuits reverse log taper or reverse reciprocal taper is desirable. Figure 1.9 shows a reciprocal taper control. It is obvious that if $R_T = F(\theta)$, then $R'_T = R - R_T$ and $\theta = 300° - \theta'$. Therefore $R'_T = R - F(\theta) = R - F(300° - \theta')$. Taper can be modified by adding an external shunting resistor. This procedure also changes

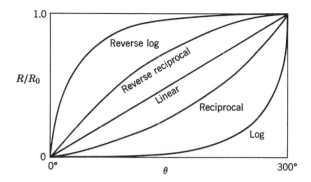

Figure 1.8 Typical carbon potentiometer taper curves.

the total resistance. Figure 1.10 shows a typical modification in which R_S is added to potentiometer R_P.

1.16   SLIDER PROBLEMS

The slider in a potentiometer provides a large source of trouble. The user should perform several tests before he accepts a potentiometer for an application. The possible areas of trouble are elucidated by the following questions:

1. Is the sliding action free of noise or bounce? When the slider is moved, does the slider stay in contact with the resistance element at all times?

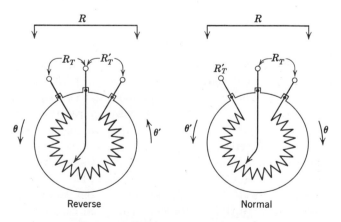

Figure 1.9 Reciprocal taper comparison.

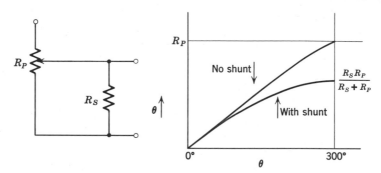

Figure 1.10   Shunt compensation of a potentiometer.

2. If the control is mechanically shocked, does the slider stay in contact with the element—over the entire element?

3. Is the sliding action or torque nearly constant over the entire control? If there are high or low torque areas, beware!

4. What is the end resistance value? How close to 0 Ω is the resistance between the slider and the potentiometer at both ends of the slider travel? In some applications a near short at one end is necessary. In some controls, 100 Ω may be the minimum value. [Mechanical designs take on two forms: (*a*) slider travel can be blocked with the slider remaining on the wire, and (*b*) an end ramp can take the slider onto a low-resistance contact.]

5. Is there hysteresis in the sliding action? A turns dial is necessary to observe this phenomenon. Make sure the control comes to the same value each time the dial is reset to the same position. Be sure the dial has no hysteresis of its own.

1.17   END EFFECTS

In some designs a tapered potentiometer is not economically practical and a multiturn control is used to obtain resolution at one end of the element. When this is done, the turns-per-inch at that one end becomes critical. It is good engineering practice to have a reserve in resolution. If the turns of wire per degree of rotation is equal to the resolution requirement, the potentiometer will just meet the need.

Controls are rated for their dissipation in two ways. One is maximum allowed current and the other is total wattage. The current rating is necessary because the wattage dissipated per unit length of the resistance is usually proportional to the resistance used. (The case considered here is the dissipation in the resistance between the slider and one end of the control.) The proper approach is to observe the current ratings given by the manufacturer. If a rating is not given, calculate the current permitted for the entire control and use this value for any part of the control. The value will be entirely safe. In high-temperature applications the derating procedures must, of course, be followed.

1.18   TAPS

At slight additional cost a tap (usually a center tap) can be added to most potentiometers. The slider action on wirewound controls should be unaffected near the tap. Any roughness near this point should be viewed with suspicion.

Consider a zero control that provides either polarity of signal for compensation as in Figure 1.11*a*. If either supply varies and the control is near center, one half of the variation appears on the slider. In Figure 1.11*b*, if the control is near center, either supply can vary and the slider will not change its potential relative to the center tap.

Center taps permit each half of the potentiometer to be treated separately. If the voltage across one half is raised, extended sensitivity on the other half is possible. If a shunt is applied to one half, the taper on that half can be significantly altered.

1.19   CAPACITORS

Capacitors are available in a wide range of types and styles. Capacitance, voltage, and size are the three factors in most capacitor selections. It is important to know how each capacitor type deviates from the ideal. To understand this requires more detailed information concerning the nature of the dielectrics. Since the dielectric dominates the specification, it is simplest to discuss capacitors in terms of dielectric characteristics.

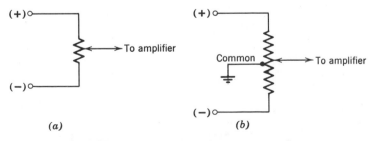

(a)   *(b)*

Figure 1.11   Use of a center tap on a potentiometer.

A dielectric material placed between the plates or foils of a capacitor determines charge-storing capability in a given geometry. The field E is just the potential applied across the plates divided by the spacing d. The displacement field D in the dielectric is equal to kE where k is the dielectric constant. The field D is directly related to the charge which is proportional to capacitance. It is desirable to use dielectric materials with high dielectric constants k, as capacitance is the ratio between stored charge and the applied voltage, that is, $C = Q/V$.

Dielectrics have many undesirable qualities. They provide leakage paths, store charge, and have ac losses. Their dielectric constants vary as a function of time and of temperature, and dielectrics break down with voltage stress. These effects are discussed by each capacitor manufacturer in his specifications. His choice of dielectric material and his care in manufacture determine the quality of his capacitor.

1.20 TERMS AND DEFINITIONS

Dielectric constant k is the ratio of charge stored in an ideal capacitor (with dielectric material present) to the charge stored if air were used as the dielectric.

Dissipation factor (DF) is defined as the ration between reactance and effective series resistance in a capacitor. This measure is frequency-dependent and for common capacitors is between 0.1 and 0.5% at 1 kHz.

Figure of merit (Q) is the reciprocal of the dissipation factor above.

Power factor (PF) is defined as the ratio

$$\frac{\text{power (in)} - \text{power (out)}}{\text{power (in)}}$$

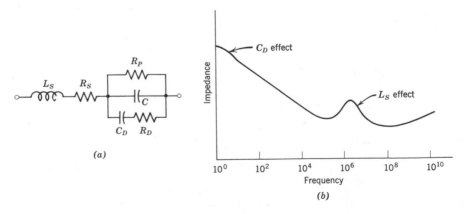

Figure 1.12 Equivalent circuit of a capacitor

This ratio reduces to the value R/Z, since for most capacitors PF $\simeq$ DF because $R << X_C$ and $Z = X_C$.

Insulation resistance is the dc resistance that permits current flow in the dielectric of a capacitor.

Frequency effects. The equivalent circuit of a capacitor in lumped parameter form is shown in Figure 1.12. The capacitor C has a dc leakage resistance R_P, a series impedance limit for high frequencies defined by L_S and R_S, and a dielectric absorption character defined by C_D and R_D. Typical values for a 0.01-μF Mylar[1] capacitor might be

$$C = 0.01 \ \mu\text{F}$$
$$L_S = 3 \ \mu\text{H}$$
$$R_S = 0.2 \ \Omega$$
$$R_P = 10^9 \ \Omega$$
$$R_D = 10^{11} \ \Omega$$
$$C_D = 20 \ \text{pF}$$

1.21   DIELECTRIC ABSORPTION

Dielectric absorption of charge in a capacitor is an often-neglected factor in capacitor selection. After a capacitor is charged to a set voltage part of the charge is captured within the dielectric and cannot be readily removed by shorting the leads of the capacitor. A discharged capacitor with dielectric absorption will slowly develop an open-circuit voltage. Any new released charge can be removed from the capacitor by reshorting the leads. After each discharge a lesser absorbed charge again produces an open-circuit voltage.

The dielectric absorption is expressed as a percentage and is the ratio of recovery voltage to charging voltage. This ratio is a function of charging time, applied voltage, discharge time, and temperature. The absorption figures that are quoted are therefore meaningful only when comparisons are made between dielectrics subjected to the same test.

The best materials are polystyrene and Teflon[2] with figures of 0.02%, polycarbonate is 0.08%, Mylar is 0.2%, mica is 0.7%, and oil-impregnated paper is 2%.

A typical test uses a 200-V charging voltage for 1 min, discharges for 2 sec, and measures the released charge in 1 min. All tests are performed at 25°C.

Dielectric absorption must be considered in the design of circuits in which time delay or discharge time is critical. Because stored energy

---

[1] Mylar is a Dupont trademark.
[2] Teflon is a Dupont trademark.

is released over seconds of time, the low-frequency capacitance appears to be greater than the midrange value. This effect is perceptible in Mylar capacitors when precise values are needed, that is, in passive filter applications. The critical capacitance values show their effect in the —3dB points, for example. The absorption difficulty can be accommodated by measuring the capacitance value at the critical frequency. This solution is economically expedient compared with the penalty of using more exotic dielectrics.

1.22   ELECTROLYTIC CAPACITORS

In electrolytic capacitors, an oxide film is used to obtain a large dielectric constant. These units have a small volume per microfarad and are used primarily in power filtering. Values range upward to 150,000 μF in aluminum electrolytics with operating dc voltages limited to about 500 V. Energy storage and capacitor volume are roughly proportional.

Ripple voltage is a measure of the reactive current flow in a filter capacitor. If the 120-Hz ripple on a 12-V supply with a 1000-μF filter capacitor is 1-V rms, the rms current would be 1.35 A. Ripple waveforms are rarely sinusoidal as nonlinear rectification processes are involved. For capacitor-input filters the peak currents in the example above could easily be 10 A if the transformer coil resistance and leakage reactance were low enough.

High ripple current can result in capacitor heating with resulting unit damage. In all designs it is wise to verify that the maximum ripple current is within acceptable limits. In capacitor-input filters a low-value series resistor can limit the peak current to acceptable limits. Since heating effects are related to the square of the current, minor limiting can be very effective.

If an electrolytic capacitor is used as a coupling element, leakage current must be considered in the design, or bias levels and operating points can be out of bounds.

Aluminum electrolytics have leakage currents given by the equation $I = 6 \times 10^{-6} \sqrt{CV}$ when I is in amperes, C is in microfarads, and V is in volts.

Tantalum capacitors have leakage currents at 25°C of 0.02 μA/μF per volt applied. Leakage current in all electrolytics increases with temperature. At 85°C, the leakage current for tantalum and aluminum is four times the 25°C value. Dry-anode tantalum is worse in this respect and increases by a factor of 10 over the 25°C value.

Tantalum capacitors are limited in capacitance and voltage range. Wet-anode foil units are limited to 1250 μF and 150 V. Dry-anode tantalum units are limited to about 350 μF and 35 V. These capacitors find application when low-temperature capacitance is important. All electrolytics drop off at −55°C from 12 to 60%. The dry-anode type is limited to about 12%. Aluminum is limited to 85°C operation, whereas tantalum can function up to 125°C. Tantalum electrolytics have a dissipation factor lower than that for aluminum. The factor varies as a function of voltage, frequency, and capacitance. Typical values for tantalum are 10%. Aluminum can run as high as 35%.

The reactance of an electrolytic capacitor reduces with increasing frequency, but most units are limited to above 1 Ω. At even higher frequencies, most capacitors look inductive and expected filtering action is not present. In these cases it may be necessary to parallel the electrolytic capacitor with another capacitor. This addition must be treated properly, as the added capacitor can resonate with the inductance of the electrolytic and a peak in impedance can result. To form a wide-band filter when an electrolytic capacitor is included requires a treatment of the total equivalent network.

When an electrolytic capacitor is used as a coupling element or to provide ac gain (as an emitter bypass), the capacitor acts as a short circuit at frequencies whose periods are greater than the RC time constant involved. The correct R value must be used in a time-constant calculation. In the case of an emitter bypass, the active source resistance of the emitter is the correct value for R. In an emitter-to-base coupling element, the dominant resistance is usually the input impedance of the base circuit.

Capacitance values for electrolytics are dependent on dc voltage, ac voltage, and frequency levels. It is undesirable to design equipment around accurate capacitance values for electrolytics. These values may be repeatable and usable for one manufacturing run of elements, but a second batch or a different manufacturer's units will not provide equivalent results.

The above statements imply that signal-related electrolytics should be used with caution.[1] In a feedback amplifier the dominant time constants should be nonelectrolytic if possible. Other time constants should be large enough so that at the lowest frequencies of interest the capacitor appears as a short circuit.

Electrolytics are usually manufactured to be out of tolerance on the high side. Tolerances of +50 to 0% are not untypical. If electrolytics

---

[1] Zener diodes are often used effectively in place of electrolytics. A Zener diode has a low impedance at all low frequencies. If the current is available to keep the Zener diode "fired," the difficulties mentioned above can be avoided.

are used below their rated voltage, capacitance values are apt to be even higher. The measure of capacitance is clouded by the standard frequencies available for measurement in most bridges. If a capacitor is to be used in a critical application, it should be measured at the critical-application frequency. The capacitance measured may not be too meaningful, but it is at least a point of useful comparison.

Dielectric absorption is present in all electrolytic capacitors. For this reason low-frequency signals can be badly distorted by signals reappearing over *portions* of each cycle. This nonlinear effect can be reduced by feedback if it is available. In open-ended applications involving filtering or decoupling, the results can be very undesirable.

Polarized capacitors can be used in a nonpolarized sense over a limited dc voltage range. Under this condition electrolytics will pass high-frequency current without damage. If bipolar operating voltages are expected, a nonpolarized capacitor is recommended.

Electrolytic capacitors are on the average larger than other nearby associated components. Parasitic capacitances to nearby points can be troublesome if the capacitor is floating and is used to couple signals. Shielding and component placement must be considered in any circuit layout. (See comments in Section 3.4 concerning parasitics.)

1.23   FILM CAPACITORS

Mylar, polystyrene, Teflon, and polycarbonate, singly, in combined forms, or in metalized form, constitute a large class of dielectric materials. The selection of one material over another is usually dictated by price and size. Often, when tighter specifications are required, polystyrene or Teflon dielectrics are necessary.

Mylar capacitors have wide acceptance for general-purpose instrumentation work. The more compact large-valued Mylar capacitors are usually metalized, although this fact is not apparent from any marking or stamping on the outside jacket or cover.

Mylar has a dielectric absorption of 0.5% at room temperature. At the extremes of temperature it increases to 1.5%. This value compares with that for Teflon and polystyrene of 0.04% over the temperature range -55 to $125°C$. Mylar has a temperature coefficient above $65°C$ of $0.3\%/°C$. A typical value for Teflon-film capacitors is 0.02% in this same temperature range. If the temperature coefficient and dielectric absorption characteristic are not too important, then Mylar provides an inexpensive, compact capacitor.

Film capacitors can be microphonic; that is, they self-generate voltages when mechanically stressed. The generating process can either be frictional or result from a change in capacitance when charges are stored in the capacitor. A well-constructed capacitor minimizes this effect.

Tab construction consists of inserting the connecting lead between two segments of foil. This construction increases the high-frequency inductance of the capacitor. In extended foil construction the electrodes are extended beyond the body of the capacitor and are soldered or welded together for a connection to terminals. This construction is preferred but it does make the capacitor larger.

Metalized-film capacitors have a distinct size advantage over plain foil units for capacitor values greater than 0.1 μF. The thin films, however, are subject to small voltage punctures through the foil. These punctures require about 10 μW-sec to clear. The resulting signal pulses can be considered noise in an analog system or a bit in a digital system. If the clearing pulse noise causes an error, then a metalized capacitor is not recommended. In some high-impedance or low-voltage applications clearing energy may not be available.

PROBLEMS

1. Noise power is proportional to the square of noise voltage, and it is additive. Consider parallel resistors of equal value. Show that the total noise voltage is reduced by $\sqrt{2}$, not 2, by considering the noise current paths for each resistor separately.

2. The resistance of a certain alloy is given by

$$R = R_0 + 2 \times 10^{-4}R_0T - 2 \times 10^{-6}R_0T^2$$

where T is temperature in degrees centigrade. At what temperature, other than $T = 0$, is the temperature coefficient zero?

3. If the temperature coefficient of copper is 0.004%/°C, what percentage change in resistance in a transformer primary corresponds to a 40°C temperature rise?

4. A wirewound resistor of 10 Ω has 1 μH inductance. At what frequency is the impedance increased by 0.5%? What should the inductance be to meet this condition at 1 MHz?

5. If 5-Ω resistors have 0.5 μH inductance and two resistors are used to make up 10 Ω, at what frequency is the impedance increased by 1%?

6. A 10-A shunt is to dissipate less than 0.1 W and be accurate to 0.1%. If copper bus has 0.1 Ω/1000 ft and is used to connect to the shunt, how accurately must the sensing points be placed to obtain this accuracy?

7. In the attenuator in Figure 1.5 C_R is 1 pF. How large must the value of C be to compensate for the attenuator? Is the compensation good at all frequencies?

8. Assume resistors of the same physical size have the same shunt capacitance. What is the improvement in effective capacitance when two equal resistors are placed in series?

9. A 10-kΩ potentiometer is placed across a low-impedance source. What is the highest source impedance found between the slider and one side of the

potentiometer? At what frequency will a 1000-pF cable load reduce the signal by 3 dB? What load resistor will affect linearity by 0.1%?

10. A 1-μF capacitor is to hold a charge to within 1% for 1 hr. What should the leakage resistance be to realize this figure? (Assume a constant charge leakage rate.)

11. What is the reactance of a 0.01-μF capacitor at 10 MHz? What inductance value resonates with this capacitance at 30 MHz? How low should the foil resistance be to equal the reactance of the capacitance at 10 MHz?

2

Magnetic Components

2.1 TRANSFORMERS

Since most electronic devices require transformers for signal handling or for power, an understanding of them is necessary in order for a designer to make useful decisions. It is not necessary to be a transformer designer, but a familiarization with design criteria makes it possible to ask for practical performance, to criticize failings, and to praise good workmanship. This section provides some theory and discusses a few of the dangers and pitfalls that can be encountered.

The core is present in a transformer for one reason: it reduces the magnetizing current to practical limits. Consider two closely coupled coils that are wound in air. The flux B and the sinusoidal 60-Hz voltage E are related by the equation

$$E = 4.44 \times 10^{-8} BnfA \tag{1}$$

This equation results from the more fundamental statement that

$$E = nA \frac{dB}{dt} \tag{2}$$

where A is the area of the coil with n turns and f is frequency in hertz. These equations do not mention a core. Once the field B is established, a similar equation relates voltage and turns to the second coil. The trouble with an air transformer at low frequencies is that the amperage demanded from the voltage source is very high. The important point here is that *transformer action still takes place providing the current is available.*

Large source currents are undesirable and a core provides the remedy. The field B in a magnetic circuit still requires a source current, but the current level is reduced by the permeability of the magnetic path. If 10 A is required in air, a core with a permeability of 10^4 would require only 1 mA.

This fact provides the answer to a key question: If the permeability varies nonlinearly with voltage or if the magnetizing current is nonsinusoidal, will the voltage waveform on the secondary be the same as the voltage on the primary? The answer has already been given. In a good transformer, yes. If the primary voltage determines B, the secondary voltage is a replica of B. When is the above statement false? If the primary coil has too much resistance, then this resistance is in effect in series with the source. Then the nonsinusoidal current demanded by the core for magnetization will modify the voltage appearing across the ideal coil. The modified voltage results in a nonsinusoidal flux B and in a modified secondary voltage.

2.2 *BH* CURVES

Even though BH curves are familiar to all engineers, a slightly different discussion of these curves is given here in the hope that a better understanding of magnetic processes will result. Figure 2.1a shows a core with a coil and a battery used to drive the coil. The BH curve of Figure 2.1a shows the result immediately after connecting the battery to the coil. The B value represents the magnetic induction of the core and the H value the ampere-turns or magnetomotive force establishing the B field. At the moment of connection, no B field exists and H rises to its starting value. If the battery is left connected for a time t_2, the B field will increase and the ampere-turns forcing the change will rise slightly (Figure 2.2b). When the battery switch is opened at time t_2, the magnetizing force is removed and the field B remains near its last value as shown in Figure 2.3b. If the switch is left open until a time t_4, the field B remains unchanged during this time as shown in Figure 2.4b. When the switch is closed again, the value of B continues to increase. The H value becomes large at t_5 but B *continues to increase linearly*. (See Figure 2.5.) After core saturation at t_7, the switch is opened.

If the battery is reversed at time t_8, the field B will return to zero and eventually reverse to the opposite saturation point at time t_{10} as shown in Figure 2.6. Note the flux slope dB/dt is constant whenever the battery is connected. This is simply Equation (2) in graphic form. If the battery is replaced by a half sinusoid of duration t_1, the BH time relationship appears to be slightly different. The B curve is

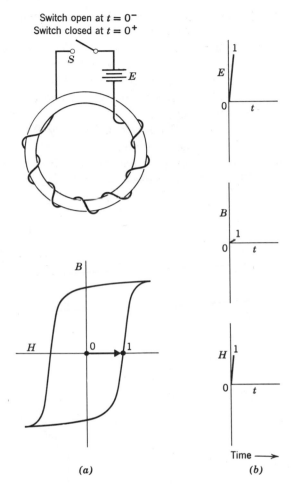

Figure 2.1   *BH* response, initial conditions.

a segment of a cosine wave. The *H* curve still has a high value at point **2**, (see Figure **2.7**). If a full cycle of a sinusoidal voltage is impressed on the core, the *BH* time relationship is that shown in Figure **2.8**.

It would appear that this transformer functions only on one half of the *BH* curve. In practice, the excess current drawn at point **2** in Figure **2.8** causes a small voltage drop in the coil and the top of the *B* curve is slightly flattened. On the return of *B* toward zero no excess current flows and the flux returns to a slightly negative value. This process repeats each cycle until finally the core operates with equal ± excursions of *B*, that is, until minor distortions in *B* are symmetrical.

The voltage level for Figure 2.8 was just large enough to saturate the core from its zero flux state. If the voltage is capable of causing saturation in both flux directions, the starting effect just described will quickly force a symmetric B excursion. A full saturating sinusoid and the resulting H values are shown in Figure 2.9. (The saturation of cores on initial turn-on is very typical in power transformers and usually requires the use of slo-blo fuses.)

Note that the B field lags the input voltage by $90°$. Since H and ampere-turns are synonymous, nonlinear voltage waveforms such as those in Figure 2.9 result from core saturation. (Note that the points of maximum current occur just before the zero crossing of voltage.)

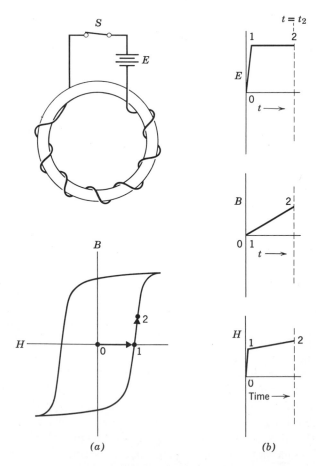

Figure 2.2   *BH* response up to time t_2.

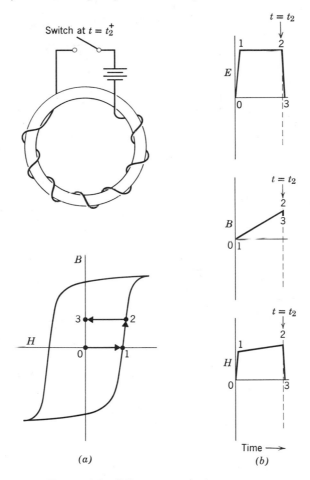

Figure 2.3   *BH* response at time $t = t_2{}^+$.

This wave form can be sensed only if the current causes an IR drop to appear somewhere in the circuit being considered.

2.3   TRANSFORMER COUPLING AND LEAKAGE INDUCTANCE

When two coils are wound on one core, the flux field B is the same for both coils. The relationship between flux B and voltage E for n turns is given in Equation 2 above, which is repeated here:

$$E = nA \frac{dB}{dt}$$

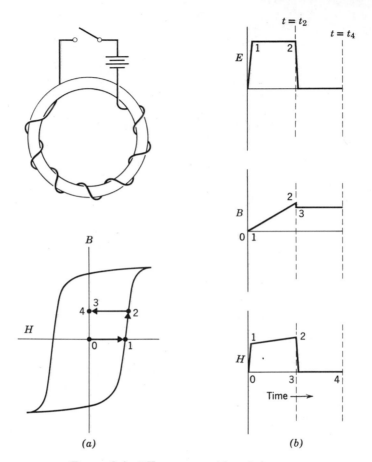

(a)   (b)

Figure 2.4 *BH* response with switch open.

The equation states that if a flux B is looped by n turns, then a voltage E results. If the second coil has $2n$ turns, the voltage on the second coil is $2E$. If the flux fails to couple a few turns, the resultant voltage is reduced.

When a resistor is placed across the second coil, a resulting current must flow by Ohm's law. This current represents ampere-turns and is in a direction to reduce the field B. The field B will prevail if the voltage source on the first coil can supply the additional ampere-turn demand of the second coil. (The first coil cannot tell whether the ampere-turns are required by the second coil or by the core.) The same effect occurs for any number of coils added to the transformer. The primary ampere-turns equal all the secondary ampere-turns plus the ampere-turns required to excite the core.

The flux pattern in a transformer with loads is not simple as the current flowing in each conductor contributes to the total field. Figure 2.10 shows a highly idealized simple transformer with arrows showing flux directions for the primary and the secondary currents. The secondary coil tries to reduce the field B in the core established by the primary coil. Since any field B must be continuous, that is, it must start and end on itself, a part of the B field leaves the core and goes around the outside of the coil. This process becomes clearer times higher for this thinner material.

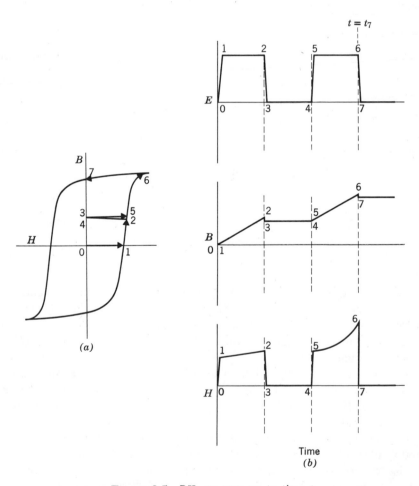

Figure 2.5   *BH* response up to time t_7.

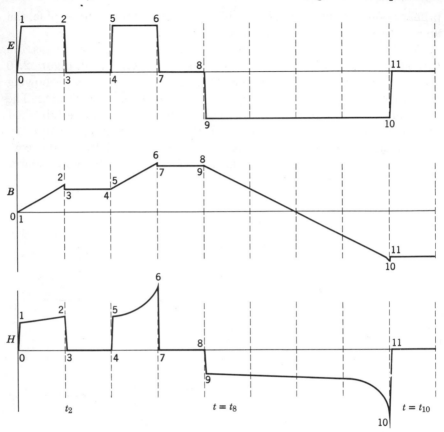

Figure 2.6 *BH* response with voltage reversal.

can exist in the secondary loop as this would cause an infinite current. Since a finite current can exist, a counter magnetic field will be set up by the second loop so that no net flux passes through the coil. The flux *B* therefore leaves the core and must go around the coil. This effect is shown in Figure 2.11.

The ampere-turns supplied by the primary does not become infinite because of the short circuit. The flux now travels in the air space, bypassing the secondary coil, and this path requires a large magnetizing current as discussed earlier.

For load currents intermediate between an open circuit and a short circuit, part of the flux established by the primary voltage will not couple to the secondary. Such a flux is called leakage flux, and the circuit parameter which measures it is called leakage inductance.

To appreciate further the idea of leakage inductance consider a shorted turn loosely coupled to the core. The field B' of this shorted turn for a given loop current is inversely proportional to loop area. The flux that must leave the core to provide zero net flux crossing of this loop also falls off as the area of the loop increases. The flux in the magnetic circuit thus divides between the preferred path and the air path in a manner that is dictated by geometry (see Figure 2.12).

The equivalent circuit for leakage inductance must provide for a division of voltages that results from a division of flux paths. Since any coil wound on a core can be used as a secondary, a leakage inductance must be associated with every coil in the transformer.

The equivalent circuit for a transformer assuming no resistive losses

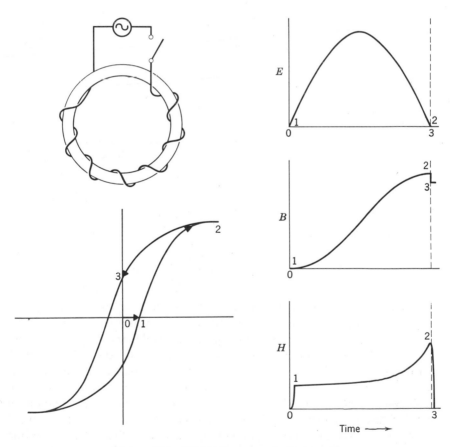

Figure 2.7 BH response to a half sinusoid.

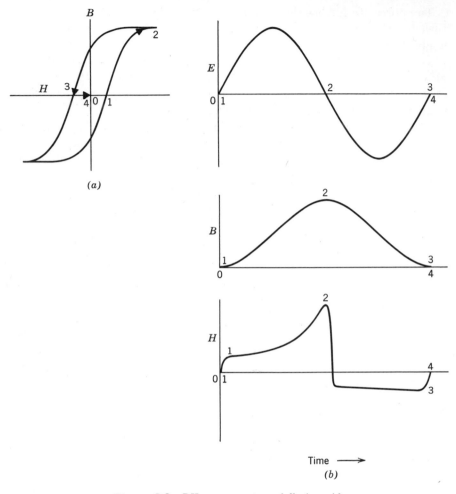

Figure 2.8 *BH* response to a full sinusoid.

is shown in Figure 2.13. The ideal segment of this equivalent circuit multiplies voltages on the primary by n, the turns ratio. The leakage inductances for the primary and secondary coils are L_1 and L_2 and are series elements, and L_m is the magnetizing inductance. The current flowing in L_m is just the ampere-turns required to excite the core. If power is to be transferred, the load current is usually much larger than the current flowing in L_m. For a short-circuited secondary, the maximum current that can flow is limited by L_1 and L_2. Here all the voltage is dropped across the leakage inductance (all of the flux avoids coupling to the secondary coil), and no voltage is left for the short circuit.

For normal loads, the leakage inductance should be low enough that voltage attenuation does not result; for example, in a typical 10-V 10-A transformer operating at 60 Hz, a leakage inductance of 10 mH causes a reactive voltage drop of 0.1 V. To keep the magnetizing current below 0.1 A, a magnetizing inductance of 0.25 H is adequate.

Coil geometry and leakage inductance go hand-in-hand. The idea of flux avoiding a turn to take an alternate path is the key to reducing leakage inductance. If the alternate paths require the flux to thread

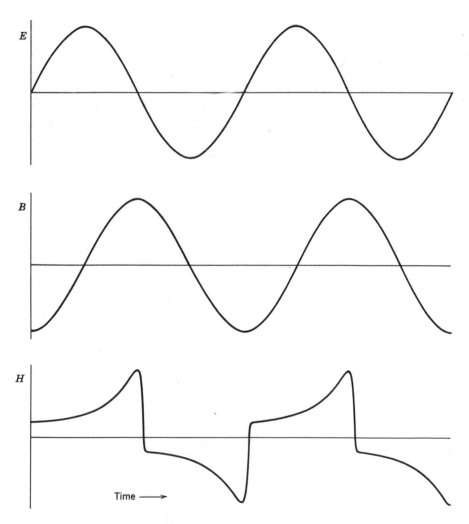

Figure 2.9   Typical *BH* response to a continuous sinusoidal input.

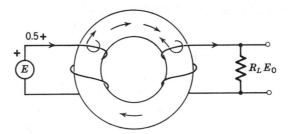

Figure 2.10 Flux directions in a transformer.

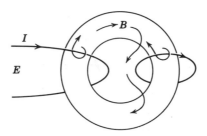

Figure 2.11 Flux pattern with a shorted turn.

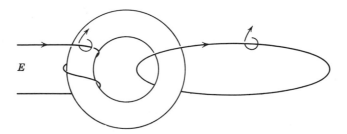

Figure 2.12 A leakage flux geometry.

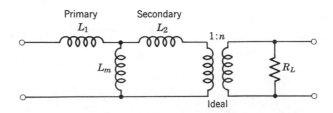

Figure 2.13 A transformer equivalent circuit.

another turn, the leakage inductance is low. If air paths are available for flux to avoid turns, the leakage inductance will be high. A few examples with toroidal cores will illustrate the relationship of geometry to leakage inductance (see Figure 2.14). (It is convenient to use toroids for descriptive purposes, but the same ideas apply to laminated magnetic structures as well.)

Typical measures to reduce leakage inductance are the following:

1. Wind coils with heavier wire or with multiple strands.
2. Wind coils so that the wire covers and circles the core several times.
3. Wind the primary coil tightly against the core.
4. Wind the secondary coil without added insulation directly onto the primary coil.

These measures have the following drawbacks:

1. Primary and secondary shunt capacitances are large.
2. Coil-to-coil capacitance is high.
3. Insulation for high voltage between coils is not provided.

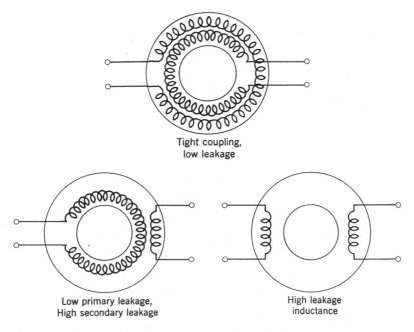

Tight coupling,
low leakage

Low primary leakage,
High secondary leakage

High leakage
inductance

Figure 2.14 Leakage inductance configurations.

4. Insulation for high voltages end-to-end on one coil is not provided.
5. Shielding is not included.

Because of cost, toroids are not used if lamination-type construction will do. The common technique is to use thin magnetic laminations cut into an E-I, F, or D-type configuration. The core is then stacked and assembled into the coils after they are wound onto a bobbin. This geometry usually does not permit the coils to cover all legs of the core. The bobbin also places an air space between the first coil and the core. Both effects raise the leakage inductance.

To use the window area of a laminated D core, effectively coils can be paralleled and located on opposite legs of the core, as shown in Figure 2.15. Note that both primary and secondary coils are paralleled. If the secondary coils are used separately (i.e., for two secondary circuits), the leakage inductance to each coil would be high. If one of the coil pairs is series-connected, the leakage inductance is low, provided the entire coil is used at one time.

2.4   A RECTIFIER PROBLEM

In a full-wave center-tapped rectifier circuit, the center point of the transformer secondary coil is brought out (Figure 2.16). The current flows during that portion of the cycle when a diode is forward-biased. Thus only one-half of the secondary is used at a time. It should be apparent that this scheme of rectification makes poor use of the copper in the transformer. If each section of the secondary is not coupled to the entire primary coil, the leakage inductance will be high. The peak current flowing into the filter capacitor is often an order of magnitude greater than the average load current. For this reason a high leakage inductance is undesirable, as it limits the peak current flow. Figure 2.17 shows a correct and an incorrect coil geometry for this case.

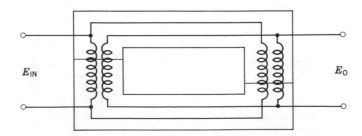

Figure 2.15 Coil locations on a laminated core (typical).

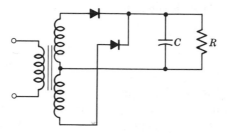

Figure 2.16   A full-wave center-tapped rectifier circuit.

2.5   DIRECT CURRENT IN TRANSFORMER COILS

The coercive force H required to saturate most core materials is small. Typically 0.1 oersted is sufficient. The relationship between coercive force and coil current is simply

$$\oint_l H \, dl = 0.4\pi n I \tag{3}$$

where l is the path length in centimeters and nI is ampere-turns. In 10 cm of toroidal core material, the maximum magnetizing force is $Hl/0.4\pi = 0.795$ ampere-turn. For 1000 turns this represents direct current of 0.795 mA.

If direct current is to flow either by design or unintentionally from element unbalance in push-pull stages, the core must have an air gap. In the example given above, an air gap of 0.01 cm would permit a much higher direct current to flow.

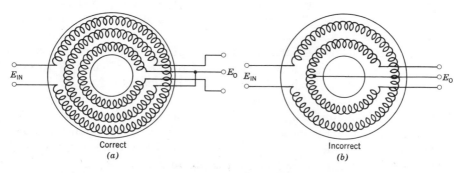

Figure 2.17   Leakage inductance geometry for FWCT operation.

Since H is equal to B in the gap and assuming $B = 10,000$ G, then

$$nI = \frac{H \times 0.01}{0.4\pi} = \frac{100}{0.4\pi} = 79.5 \text{ ampere-turns}$$

For 1000 turns this is 79.5 mA, a considerable increase over the value 0.795 mA permitted without the gap.

The presence of an air gap in any transformer reduces the magnetizing inductance and raises the magnetizing current. This is a penalty that must be paid if direct current is to flow in the coils.

C cores are cut and lapped to provide for a minimum gap. If a gap is required, a C core provides a means for setting an accurate gap. E-I laminations when interleaved provide less gap than when E's and I's are stacked together. The latter construction provides gap, but the gap thickness is not too well controlled. In most power-transformer designs this is not too critical.

Leakage inductance is essentially unaffected by core saturation as most leakage flux exists in air. Thus the equivalent circuit shown in Figure 2.13 is still correct except the effective value of L_m must change for any direct current flow.

2.6   PERMEABILITY

A single BH curve for a magnetic material illustrates only part of the relationship between B and H. The ratio between these two values is broadly called the permeability. This is an oversimplification of facts and the engineer should understand the subtleties that are hidden in this definition.

Figure 2.9 illustrates clearly that H is nonsinusoidal for a sinusoidal B. The ratio B/H can be defined in terms of peak or rms values. Peak values are usually used with the voltage held sinusoidal. It is also possible to excite the core with a sinusoidal H (current) and measure B by observing voltage. The permeability measured this way will roughly correspond to the first method.

Minor BH loops (excursions of small peak B) vary as a function of flux level, and each loop yields a different ratio of peak B to peak H as illustrated in Figure 2.18. Thus permeability is a function of peak B. If the frequency is increased and the peak flux level is held constant, the magnetizing force increases, thus reducing the permeability. This effect is shown in Figure 2.19. The thinner the core material, the higher the permeability for a given frequency of operation. Core materials with thicknesses down to $\frac{1}{8}$ mil are avialable in toroids. In laminations, 4 mils is the usual lower limit. Because of their gap, powdered cores have

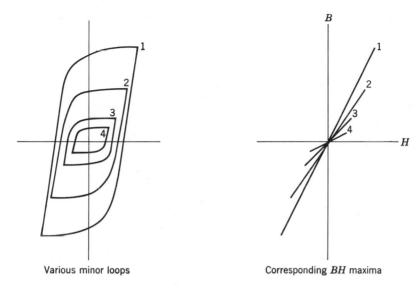

Various minor loops   Corresponding *BH* maxima

Figure 2.18 A typical BH curve and related permeability.

low permeabilities, but these permeabilities are often greater than those available in laminated materials at the same frequency. Data on permeability as a function of peak B and frequency for various laminations are thoroughly covered by the core manufacturers.

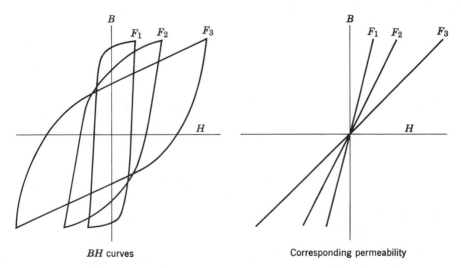

BH curves   Corresponding permeability

Figure 2.19 Permeability as a function of frequency.

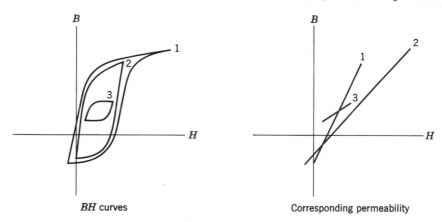

BH curves   Corresponding permeability

Figure 2.20 Permeability with added direct current.

The minor *BH* loops described above assumed no dc flow. The permeability is usually reduced by the presence of direct current, as illustrated in Figure 2.20.

2.7 CORE LOSS AND EQUIVALENT CIRCUIT

The permeability figure for a core provides only one measure of core performance. Core loss per pound of material as a function of frequency and as a function of maximum *B* are also necessary data for a design. Core loss usually dominates at high frequencies and implies that the shunt impedance in the transformer equivalent circuit must include resistance. The equivalent circuit for a transformer including core loss, coil resistance, and shunt capacitance is shown in Figure 2.21.[1] L_m and R_f

[1] To refer elements across the turns ratio, multiply impedances by n^2. If $n = 10$, then $L_s = 1$ μH becomes 100 μH, $C_s = 100$ pF becomes 1 pF, and $R_s = 1000$ Ω becomes 100 kΩ.

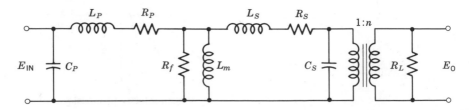

Figure 2.21 An equivalent circuit for a transformer, showing losses.

can be approximated from the permeability and core-loss curves given
for each type of material.

2.8   TRANSFORMER ELECTROSTATIC SHIELDS

Electrostatic shields in a transformer are a necessary part of many
designs. Their criticality varies significantly, with very "tight" shields
adding to the cost. Often two or more shields are required and the engineer should know how to sort them out, how to specify them, and what
to do with them.

To understand the use of an electrostatic shield in a transformer,
it is necessary to understand the electrostatic-shielding process in general.
Electronic circuitry is usually surrounded by a metal housing. This housing is usually connected to the zero reference potential of the signals
being processed by the circuit. When so connected, the housing has capacitances to all points in the circuit. Since these capacitances terminate
on points of zero reference potential, no unwanted signals can couple
into the circuitry. The housing is usually called a shield.

A transformer provides convenient power inside an electrostatic enclosure. When the secondary coil is connected to the circuitry and the
primary is connected to the utility outlet, the primary-to-secondary coil
capacitance violates the shield integrity. This capacitance, rather than
terminating on a zero reference potential, terminates on an unknown
ground potential in series with some of the primary coil voltage. Such
a circuit is shown in Figure 2.22.

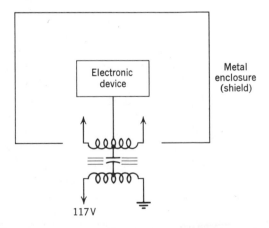

Figure 2.22   A shield enclosure and a transformer.

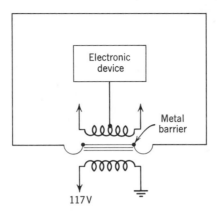

Figure 2.23 A transformer shield used to maintain a tight total shield around some electronics.

To close such an opening in the shield enclosure, an electrostatic barrier can be placed between the coils of the transformer. The barrier is usually a sheet of thin copper foil wrapped around the primary coil. Since the foil is actually one turn, care must be taken to keep the ends from touching and causing a shorted turn. The foil has *no influence* on the magnetic processes in the transformer. It does add to the space between the primary and secondary coils and also adds to the coil capacitances. The added piece of metal foil can be connected to the housing, thus closing the hole in the electrostatic enclosure. This technique is shown in Figure 2.23. Such a metal barrier is correctly called a transformer shield. The shield connection shown in this figure has a drawback that relates to the flow of reactive energy in the coil-to-shield capacitance. This current can flow in conductors that carry signal potential differences. Often it must be held below microamperes to protect low-level signals. A typical shield problem is shown in Figure 2.24 where current flows in the loop ① ② ③ ④ ⑤ ⑥ ⑦ ① and in particular in conductor ⑥ ⑦. Here coil voltage ① ② causes reactive current to flow in capacitance C_{23}.[1]

To eliminate this particular loop, the shield ③ could be connected to ① so that the reactive current would now flow in the loop ① ② ③ ①. When this is done, primary-to-shield capacitances affect the circuit as shown in Figure 2.25. Current flows in the loop ① ② ③ ④ ⑤ ⑥ ① and in particular flows in conductor ④ ⑤. Here coil voltage ① ② and

[1] The transformer has an iron core but for clarity the iron designators have been omitted from the drawing.

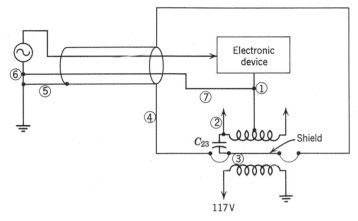

Figure 2.24 Current-flow transformer shield connected to local shield.

ground difference of potential ① ⑥ circulate current in capacitance C_{23}. Proper current-flow control can be obtained only by the use of two shields. This shield configuration is supplied with most low-level instruments. The second shield can be connected to utility ground but this does not eliminate the effects of ground-potential difference ① ⑥. A best solution is to connect the primary shield to signal-source ground coaxial with the input shield[1] as shown in Figure 2.26. Coil currents

[1] The input cable shield can be used to return the primary shield if required. This reduces flux capture in the loop ⑨ ⑩ ④ ③ ⑧ ⑨ ⑩. Any flux crossing this loop causes current to flow in conductor ④ ⑩.

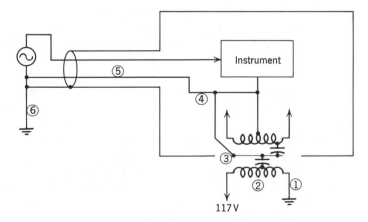

Figure 2.25 Current-flow transformer shield connected to signal common.

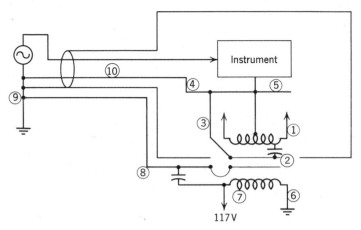

Figure 2.26 A proper double-shield transformer connection.

flow in loops ① ② ③ ④ ⑤ and ⑥ ⑦ ⑧ ⑨ ⑥, but these currents do not flow in a signal conductor.[1]

The foregoing discussion provides three distinct uses of shields in transformers:

1. To complete the electrostatic closure around a circuit.
2. To control the flow of unwanted transformer currents.
3. To provide a path for the flow of unwanted external currents.

2.9   LEAKAGE CAPACITANCES

The shields in a transformer can be measured for leakage capacitance to determine their effectiveness. The leakage capacitance is measured by driving all coils within a shield with respect to the shield and observing pickup on all other conductors. This test is shown in Figure 2.27. Typical generator settings are 10 V and 1 kHz. A pickup of 1 mV implies that the capacitive reactance is 10^8 Ω or the capacitance is 1.6 pF. It is *imperative* that the generator lead be well shielded to obtain useful results.

In a 20-W 60-Hz power transformer, shields that are made up of single turns of copper strip have typical leakage capacitances of 20 to 100 pF. Shields that are pinched over a coil on the edges with shielded wire brought out to the header have typical leakage capacitances of

[1] The reader is referred to the author's book *Grounding and Shielding Techniques in Instrumentation*, John Wiley & Sons, 1967, Chapter 4.

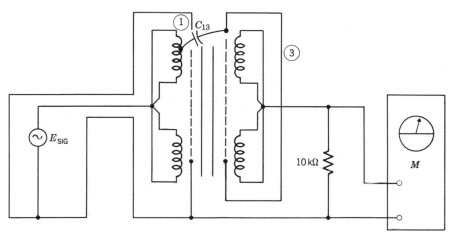

Figure 2.27 A measurement of shield effectiveness.

3 to 10 pF. Box shields, that is, shields that carefully cover each coil, have typical leakage capacitances of 0.1 to 1 pF.

The need for tightness in a shield varies with the application. The simplest analysis assumes that a capacitance exists from each coil to a point immediately outside its shield. The coil voltage in series with this capacitance causes a loop current to flow, returning to the point of zero voltage on the coil. The current produces potential drops in conductors or elements in the loop that will affect circuit performance. To determine the required leakage capacitance of the shield, calculate the effect of the current and deduce the largest capacitance permitted.

2.10 MULTIPLE SHIELDS

Two and three tight shields are often used in instrumentation transformers. An example shows how three shields might be connected. Consider a current feedback amplifier powered from a 60-Hz power transformer. The circuit is shown in Figure 2.28. Note that a feedback resistor R goes between signal common and amplifier common. To eliminate unwanted current flow in R from the secondary coil a secondary shield is required. To guard against primary ground current flow, the signal electrostatic shield is present (center shield). The primary shield keeps induced primary currents from flowing in the input-signal conductor.

The foregoing circuits show single-point signal grounding as only the transformer shields are being discussed. The problems of multiple-point signal grounding are discussed in Chapter 7.

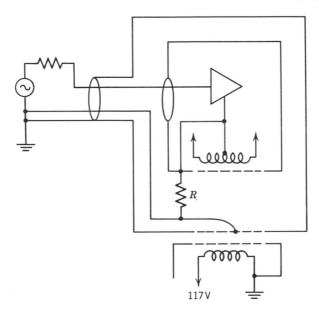

Figure 2.28 A three-shield transformer.

2.11 PRACTICAL SHIELDS

Toroids are difficult to multiple shield. This is particularly true of small toroids. The bunching of the shield material in the center makes the job tedious and messy. Two advantages do exist however. The inner coil does not have to be shielded from the core if the coil covers the core, and the last shield does not have to thread the hole and can be wrapped outside the entire toroid. The metal case used to house the toroid can also act as the outer shield.

Toroids are often shielded by wrapping the coils with a narrow strip of metal-backed insulation. The edge of the strip is insulated to avoid shorted turns. This type of shield is effective as an electrostatic barrier, but it can couple transformer voltages because the shield is itself a multiturn coil. When one point of the shield is connected to a zero of potential, other points on the shield are at a potential determined by the volts-per-turn for the transformer. The higher the volts-per-turn, the more ineffective the shield.

Stacked cores with interleaved coils and shields are the simplest to construct. If the primary is split on two legs, the crossing of shielded conductors to protect the shield integrity becomes complicated. For this reason shielded interleaved split coils are not too practical.

For very tight or "box" shields the shields and coils are often preformed as an assembly with leads included. Such a structure can be slipped into place around another coil or onto the core as required. Although this practice is somewhat wasteful of window area, it does provide for excellent shielding.

The housing around a transformer can be connected electrically to the core or not as required. In designs without proper shields, core and case connections are sometimes critical. If the core is one side of a shield, it should be connected accordingly. No specific recommendations can be made in the case of the unshielded transformer. The user will have to experiment for himself for best results.

If a header is used to terminate all coil leads and shields, shield effectiveness will be reduced at the header. It is customary to dress shielded wires out through holes in the transformer case to avoid disturbing the shielding. The negative aspects of this procedure are (a) shielded wires are more difficult to handle, (b) holes for shield exits reduce the effectiveness of the case as a magnetic shield, and (c) hermetic sealing is not possible.

All shield segments, splices, and ties inside the transformer should be solder connected to avoid noise and intermittency problems. This requirement adds to the cost of construction, but it is necessary for good performance. The user should be aware of the problem and should test transformers for noise and intermittency during application by hitting or tapping on the case.

Conductive paper can be used as a transformer shield. Shorted shield turns do not affect transformer operation and the power requirements are minimal. Typical resistivity figures might be 200 Ω per square. A 200-Ω shield turn with 0.2 V per turn would dissipate only 0.2 mW. Connections to the paper can be made by a gummed-back copper strip.

Shields are usually involved in high-impedance circuits. Leakage capacitances are often less than 20 pF, and at 60 Hz this is a reactance of 130 MΩ. It is apparent that a 200-Ω shield resistance will not affect the performance of the shield.

2.12 SHIELDS AND LEAKAGE FLUX

Shield-lead dress is sometimes all important. One example occurs in capacitor input filters where the diodes conduct over a small part of each cycle. This large peak current involves a corresponding change in leakage flux. The flux is coupled to the loop area involving the shield connection and the circuit connection to the coil being shielded.

The induced voltage in this loop is dependent on $d\Phi/dt$ which is maximum each time a diode disconnects from its load. It is not uncommon to pick up 1.0-V spikes of 10-μsec duration between the shield and parts of the coil.

To reduce this effect, coil center taps and shield ties should be made with minimum loop area. This phenomenon also involves shield-to-shield loops. Best practice requires that all coil and shield connections be made close together to avoid any loop area near the core. It is also good practice to use low-leakage transformer construction if the problem is critical. Current induced into a circuit by flux coupling is called an insertion current.

2.13 SHIELD TIGHTNESS

Transformer shields can be measured for leakage capacitance as indicated in Figure 2.27. The leakage capacitances can be reduced to below 0.01 pF if care is taken. Since the measurements are not made with the transformer in operation, they are only one indication of performance. For electrostatic guarding, leakage capacitance measured with the transformer unenergized is correct. For a measure of all signal coupling, the transformer must be tested in operation. The example of spikes coupled by the leakage flux or of the single-turn voltage on a shield illustrates this point. Low-leakage capacitance shields are often provided but, if the unwanted coupling is still high, the result is still improper. Further reduction of leakage capacitance does not solve the problem.

2.14 CUP CORES AND POT CORES

If the wire and core material in a toroid are interchanged, the wire can then be wound on a simple bobbin. The core wraps around the coil and completely encloses the coil. Since the core material has a distributed gap, the segmenting of the core for removal adds to the uncertainty of the gap dimension. Cores are available with variable air gaps for tuned inductor applications. Materials in this configuration are called cup cores.

Cup cores provide an inexpensive bobbin method of winding transformers or inductors. Since the core surrounds the coil, a low external flux field results. This feature is called magnetic self-shielding.

2.15 EXTERNAL MAGNETIC PICKUP AND COUPLING

A coil geometry that reduces leakage inductance also tends to reduce external flux coupling. Toroidal construction is ideal as the leakage flux can be kept low. Inductors wound on toroids are less susceptible to external pickup than units wound on laminated cores.

If the mounting case of an inductor is made of magnetic material and the core has low permeability, then the case can change the magnetic path and modify the inductance. A metal case or the proximity of other coils can also influence the inductance because of mutual effects. These effects are usually second-order, but in tight tolerance work they must be considered.

2.16 MAGNETIC MATERIALS, SILICON STEEL

The bulk of magnetic material used in transformers and motors is silicon steel. The amount of silicon, the annealing process used, grain orientation, and rolling direction produce a variety of standard products. Laminations are usually available in 12- and 14-mil thicknesses.

The maximum flux density approaches 20 kG with a coercive force from 0.6 to 0.9 oersteds for 10 kG. Permeabilities vary between 4000 and 10,000 at 7 kG. At 200 G, the permeability is reduced to about 1100.

2.17 GRAIN-ORIENTED SILICON STEELS

TRADE NAMES: *Orthosil, Hipersil, Magnesil, Oriented T, Selection, Microsil.*

Coil wound tapes are common forms for this material. The layers are insulated to reduce eddy current losses and bonded together to yield a rigid core. The cores are then cut and the surfaces carefully polished. Halves are supplied mated so that minimum air gap will result when the cores are placed together. Halves are tightly banded together after assembly through the coil form.

The material has good permeability and high peak flux density and is used extensively in power transformers. The peak induction is near 20 kG, the permeability is above 15,000 at 10 kG. For 12- to 14-mil-thick material, the coercive force is 0.1 oersted at 10 kG. Some manufacturers provide this material in standard lamination sizes. Stamping, however, reduces the peak magnetization and peak permeability.

Tape-wound toroidal cores are available with grain-oriented silicon steel in thicknesses from 1 to 6 mils. The coercive force is about five times higher for this thinner material.

2.18 HIGH-PERMEABILITY NICKEL IRON
(48% NICKEL, BALANCE IRON)

TRADE NAMES: *Alloy 48; Monimax; Hipernik; AL 4750; 49 Permalloy; Deltamax; Unimag 50; Supermu 10; 48 Alloy; 48 NI; High Permeability 49.*

This alloy has general power and audio use. The maximum flux density is 14 kG. Coercive force at 10 kG is 0.2 oersted and that at 1 kG is 0.015 oersted. The permeability maximum is 40,000 to 130,000 at 6 kG, and the incremental permeability in the coersive range 0.01 to 0.1 oersted for 1 kG incremental variation is 6000 to 13,000. Core loss is 0.2 W/lb at 60 Hz for 14-mil material at 10 kG. It is available in tapes and in laminations. If the alloy is cold-reduced in grain-oriented form, a very square hysteresis curve material can be produced. This type of magnetic material is used in saturable reactors and in magnetic amplifiers in which the permeability falls off very rapidly once the peak induction has been reached. The coercive force is nearly constant and varies between 0.2 and 0.3 oersted.

2.19 IRON COBALT AND VANADIUM

TRADE NAMES: *Vanadium Permandur;*[1] *Supermendur.*

These materials are high-induction materials. An alloy of equal amounts of iron and cobalt, and with 2% vanadium provides a peak induction of 23 kG and permeabilities with special magnetic annealing of 50,000 at 16 kG. This material is used when core size is critical and the core must be operated at a high flux density.

2.20 MOLYBDENUM NICKEL IRON

TRADE NAMES: *Permalloy 80; Mo Permalloy; 4-79 Permalloy; Supermu 30, 40; Hy-Ra 80; Unimag 0-80; Supermalloy; Moly Permalloy; Hipernom; Hypernom V; Hy Mu 80; Unimag 80.*

This alloy is about 4% molybdenum, 79% nickel, and the balance iron. It is not grain oriented. It is a basic material for powdered-iron cores, tape cores, and laminations, usually where high frequencies are involved. The alloy has excellent incremental permeability and low core

---

[1] Allegheny Ludlem trade name.

loss. In tape form, the saturation induction of this material is about 7500 G. Permeability at 200 G is high, roughly 50,000. Maximum permeability at 3000 G varies from 100,000 to 400,000. Coercive force is low and is typically 0.01 to 0.03 oersted at 3 kG. Core loss for 4-mil material at 3 Kg at 60 Hz is 0.006 W/lb. At 1 kHz, this figure rises to 0.3 W/lb.

If the heat treatment and rolling operations are carefully controlled, this alloy can exhibit phenomenally high initial permeability; for example, Supermalloy[1] has a guaranteed initial permeability of 55,000 at 20 G measured at 100 Hz. In powdered form, permeabilities vary from 3 to 250. The standard configuration is toroidal with a large selection of cross sections and window areas available.

2.21   IRON, COPPER, CHROMIUM, NICKEL, AND IRON ALLOY (MU-METAL)

A special alloy containing the above elements can be made to have superior forming characteristics, making it desirable as a magnetic shield material. To realize the full magnetic properties of this material, it must be annealed after fabrication as mechanical strain disturbs its magnetic properties.

This alloy has excellent permeability at low flux densities and should be used to shield only from outside low-level magnetic phenomena. Since it has a peak induction of only 7 kG, it should not be used near a large magnetic field, for example, to surround a power transformer.

2.22   POWDERED-IRON CORES

A variety of powdered-iron materials for cup cores, toroids, slugs, etc., are available. By varying the material composition, permeabilities are varied from 3 to 85. Low-permeability materials have frequency applications above 250 mHz. Saturation flux densities are approximately 8 kG.

In manufacture, iron in finely divided form is processed for particle insulation and then a thermo-setting binder is added. The material is then compressed into various forms under very high pressure after suitable lubricants are added.

Powdered-iron cores are not as stable as powdered-permalloy equivalents, but when cost is critical these materials find wide application. The insulation properties keep the core losses low and result in excellent high-frequency permeability. The cores have temperature coefficients Q

---

[1] Arnold Engineering Co. trade name.

and permeability both of which must be considered in applications. When magnetically shocked, these materials take a set in both permeability and Q but eventually recover.

2.23　FERRITE CORES

Manganese zinc ferrite and nickel zinc ferrites are formed and processed by ceramic techniques. Their oxide structure makes them behave as insulators with resultant excellent high-frequency operation. The manganese ferrites have permeability as high as 3400 and are useful in the low kilohertz range. The nickel ferrites are suitable for operation to 80 MHz with a permeability of 160. The saturation flux density for this material is below the powdered-iron material and runs between 1900 and 3400 kG, and its initial permeability is high and can be controlled as a function of temperature. The Curie point on these materials can be as low as 135°C which limits high-temperature applications.

2.24　INDUCTORS

Perfect inductors do not exist. With this understanding the engineer is better equipped to select an inductor for his application. To understand the details of the problem, some theory is presented.

An inductor is a device for storing energy in a magnetic field. The equation relating inductance to energy is

$$E = \tfrac{1}{2}LI^2 \tag{4}$$

where E is in joules, L is in henrys, and I is in amperes. In terms of the magnetic field this expression takes the form

$$E = \int \frac{B^2}{8\pi\mu}\,dV \tag{5}$$

where B is the field of magnetic induction in gauss, μ is the permeability in each volume element, and volume V is in cubic centimeters. Two other equations are needed: $B/H = \mu$ where H is the magnetizing force in oersteds and $\int H \cdot dl = 0.4\pi nI$ where I is the path length in centimeters and nI is ampere-turns.

Consider a closed magnetic path of high-permeability material. Let us assume that the material is to be operated at a flux density of 10 kG, that it has a permeability of 10,000, and that the volume of magnetic material is 10 cm³. The field outside of the core can be neglected, and the total magnetic energy stored at full flux density, by using (5), is

$$E = \frac{(10 \times 10^3)^2 \times 10}{8\pi \times 10^4} \text{ ergs} = 0.002 \text{ J} \tag{6}$$

The energy storage required in a power-supply filter inductor might be 0.2 J (0.1 A and 20 H). Obviously the inductor in this form can store only 1% of the energy required. The key to storing energy rests in the denominator term of (5). If the field B can exist where μ is low, the energy stored in that volume can be significant. To realize this in practice, a gap is provided in the magnetic path. Since B is nearly continuous across the gap, the field B exists in a region of unity permeability.

Let us assume a gap volume of 0.1 cm³. The energy stored in this gap for a B of 10 kG is 0.2 J, whereas the energy stored in the core is only 1% of this value. Thus it is clear that energy storage in inductors implies a gap of low-permeability material. *Energy is not stored in the core. The core makes it possible to store energy in a gap.*

It is of interest to calculate the inductance of our gapless core and coil. If we assume 8000 turns of wire and a magnetic path length of 10 cm, 8 ampere-turns would be required to excite the core to 10 kG. (The coil current in 8000 turns would be 1 mA.) The inductance of this coil can be calculated by equating the two energy statements (4) and (5). Since $\frac{1}{2}LI^2 = 0.002$ J and $I = 10^{-3}$ A, L is 4000 H. Thus very large inductance values occur without a gap, but the resulting inductance is not able to store energy.

If the gap length is 0.1 cm, 800 ampere-turns are required to support an induction field of 10 kG in the gap. (The core and the gap together actually require 808 ampere-turns.) For 8000 turns the coil current is 0.1 A and the energy equivalence shows that the inductance is 20 H.

The gap in many inductors is distributed throughout the core, for example, powdered-ferrite cores or powdered-permalloy cores. The filler material in these cores can be varied to modify their effective air gap. The preceding discussion shows very clearly that the presence of a gap in a magnetic medium greatly reduces the inductance. It is also true that a lower permeability material reduces the inductance. In a powdered core it is not convenient to discuss the gap dimensions even though the gap is present. It is simpler to discuss the resulting permeability of the material which provides equivalent design information. The equation relating inductance to permeability is

$$L = \frac{1.256n^2 \times 10^{-8}}{l_m/\mu a_m + l_a/a_a} \tag{7}$$

where l_m and l_a are lengths in centimeters of metal and air paths, a_m and a_a are cross-sectional areas of metal and air paths in square centimeters, n is turns, and μ is the permeability in metal. Powdered core material with high-permeability material has a reduced effective air gap, and the energy-storing qualities are reduced. Powdered core materials are available with permeabilities that cover the range 10 to 500. The gaps in these powdered cores are very stable and not subject to change under mechanical stress or temperature variation. For this reason powdered cores usually yield very stable inductors.

The current I in (4) is an instantaneous value. If the inductor is used only for sinusoids, the peak energy is stored twice per cycle when the peak current is $\sqrt{2}I_{rms}$. The peak energy stored per cycle for ac excitation is twice the average value.

In many inductor design problems, a direct current is superposed on the alternating current. It is important to use an inductor that will not saturate in the presence of the two signals. The peak energy storage is

$$\tfrac{1}{2}L(I_P + I_{dc})^2 \tag{8}$$

At a constant voltage level the current in an inductance varies inversely with frequency and the peak energy stored is proportional to the square of the peak current. In any design, the lowest frequency considered and the largest direct current define a largest total current. The inductor must be able to store the energy defined by this total current, namely,

$$E = \tfrac{1}{2}LI_{max}^2 \tag{9}$$

The flux density limits for each core material are described by the manufacturer. The equation relating voltage V, frequency f, and magnetic induction B is

$$V = 4.44 \times 10^{-8}BnfAb*$$

where A is the core cross-sectional area in square centimeters and n is coil turns. The voltage is the *rms value of a sinusoid*. The induction B is *peak flux density in gauss*. It is important to note that this equation is independent of (*a*) gap or gap material, (*b*) permeability, (*c*) magnetic path length, (*d*) wire size or resistance, (*e*) core material or thickness, (*f*) core geometry (except area).

---

*The factor b is a stacking factor used to modify the effective core cross-sectional area.

2.25   INDUCTANCE VERSUS FREQUENCY

Magnetic materials have permeabilities that vary with frequency and with flux level. As shown by (1), if the frequency is increased and the voltage is held constant, the peak flux level reduces. In general the permeability decreases with reduced flux level (see Section 2.6). To maintain a reasonable permeability as a function of increasing frequency the lamination thickness used in C cores or stacked cores must be reduced. Powdered cores by their very nature are usually good at high frequencies.

If the principal characteristics of an inductor are defined by the gap geometry, variations in permeability as described above have little effect on the inductance. If 99% of the energy is stored in the gap, the permeability can vary 2:1 and cause an inductance change of just 1%. If the inductor stores its energy in the magnetic material, the inductance will vary directly with permeability. Permeability can change several orders of magnitude as a function of flux level or frequency. Inductors requiring a large dynamic range or operation should have the ratio of gap energy to core energy as large as possible.

Powdered cores have permeabilities that vary slightly as a function of frequency. This information is well documented by the core manufacturers and should be considered in each design. Each core type has a preferred frequency range; for example, ferrite cores are preferred above 10 kHz, high-permeability permalloy cores make excellent medium-frequency-range inductors, and stacked 12-mil laminations with a suitable gap make excellent power inductors. All inductors have parasitic shunt capacitance. This capacitance results in a parallel resonant tank circuit. It is desirable to have the tank circuit's natural frequency as high as possible because above resonance all inductors look like capacitors. It is typical to have inductors in the henry range resonate below 1 kHz and to have inductors in the microhenry range resonate near 30 MHz.

The effective shunt capacitance can be calculated by determining the equivalent energy storage in the electric field. (This calculation is discussed in Section 1.8.) Once the electric field energy is known the effective capacitance and the energy storage are related by the expression

$$E = \tfrac{1}{2}CV^2 \tag{10}$$

It is apparent from this equation that the start and finish turns in an inductor should be well separated. In general, greater coil lengths result in higher natural frequencies as the turn separation is greater. Some inductors are pi section wound or wound in a honeycomb pattern to reduce capacitance. The important comment to be made here is that considerable variation in natural frequency is available depending on the construction technique used. It is discouraging indeed to place an inductor into a circuit only to find out later that in the frequency range of interest it is an expensive capacitor.

2.26   CORE LOSS, COPPER LOSS, AND Q

Inductors used at high frequencies usually function at low flux densities. This implies that the losses in the core per cycle are low. If an inductor is used at high flux levels, then the type and thickness of material determines the core loss per cycle. These effects are well documented for each core type. Such dissipation is termed "eddy-current" loss.

At high frequencies the wire in the coil is only partially effective because of skin effect. The ineffective copper cross section raises the resistance of the coil. One remedy for this loss is to wind inductors with Litz wire (multiple strands of insulated fine wire). Litz wire is effective in the frequency range of 30 kHz to 1 MHz and is available in many variations of strand size and strand number.

The losses in an inductor are discussed in terms of a parameter called Q. The Q of any energy-storage system is defined as

$$Q_L = \frac{2\pi \times \text{energy stored per cycle}}{\text{energy lost per cycle}} \tag{11}$$

and for an inductor this becomes

$$Q_L = \frac{\omega L}{R} \tag{12}$$

where ω is frequency in radians per second and R is an equivalent series resistor representing all losses.

Because of the effects discussed earlier, Q does not vary linearly with ω as (12) would indicate. Manufacturers provide typical Q curves for each core configuration indicating the region of frequency with maximum Q. One equivalent circuit for an inductor showing losses is presented in Figure 2.29. The elements in this circuit are as follows:

R_1 represents core loss that varies with frequency ($R_1 = k_1 f$)
R_2 is copper resistance
R_3 is copper resistance that varies with frequency ($R_3 = k_2 f$)
C is parasitic capacitance
L is an inductance that varies with frequency ($L = L_1 - k_3 f$)

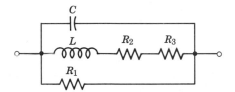

Figure 2.29   Inductor equivalent circuit.

To measure the Q of an inductor it is necessary to resonate the inductor at the frequency of interest. This can be done by shunting the inductor with a suitable capacitor. (This technique assumes that the natural
frequency of the inductor is above the frequency of interest.) At the
resonant frequency the tank circuit appears like a resistor of value
$\omega L (1 + Q)$. If the measuring circuit is that of Figure 2.30, then

$$Q = \frac{E_O}{E_{IN} - E_O} \left(\frac{R_1}{\omega L}\right) - 1 \tag{13}$$

It is sometimes difficult to know the value of L if the measurement
is made just below the natural frequency. In this case the resonant
frequency is measured with and without an added capacitance. The inductance is then

$$L = \left(\frac{1}{f_2^2} - \frac{1}{f_1^2}\right) \frac{1}{4\pi^2 C} \tag{14}$$

where C is the added capacitor and f_1 and f_2 are the resonant frequencies
with and without the added capacitor C. All of this discussion assumes
that the capacitor has a high Q, a valid assumption in most cases.

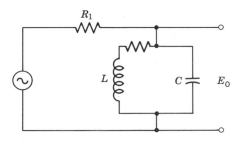

Figure 2.30   A circuit for measuring Q.

PROBLEMS

1. According to 2.1 (1), if A is the cross-sectional area of a transformer core, how much more core area is required at 50 Hz? How much less at 400 Hz?

2. A core and coil combination will just support 10-V peak-to-peak square waves at 1000 Hz. If a 10-V battery is switched across the coil and the initial B is zero, how much time will elapse before the core saturates?

3. In Problem 2, if the square-wave frequency is 500 Hz, what peak-to-peak voltage will saturate the core?

4. What is the amplitude ratio of an rms sine wave to a peak square wave at that frequency which will just saturate the same core? (The average value of a sinusoid is $0.637E_P$.)

5. In Figure 2.21 assume $R_P = R_S = C_P = C_S = 0$ and R_f and L_m open. If n is 2, $f = 1000$ Hz, and $L_P = L_S = 10$ μH, how much output signal attenuation occurs for $R_L = 100$ Ω? What value must $L_P = L_S$ be to keep the loss 10% at 10 kHz?

6. In Figure 2.26 in loop area ② ③ ④ ⑩ ⑨ ⑧ ②, if C_{2s} is 1000 pF, the loop area is 100 cm², and $B = 100$ G at 60 Hz, how much current will flow in conductor ④ ⑩?

7. In Figure 2.27, if $E = 10$ V at 20 kHz and the output is 1.0 mV, what is the value of C_{13}?

8. At what frequency is a 100-mH inductor in error by 1% as a result of 100 pF of shunt capacitance?

9. Using 2.24 (7) assume $l_m = 10$ cm, $l_a = 0.1$ cm, $\mu = 5000$, and $n = 1000$. Calculate the resulting inductance. Can the dc energy-storage capability be determined by this equation?

10. Two inductors of 1 mH with Q's of 100 are paralleled. Calculate the resulting Q. What is the series Q value?

3

Parasitic Effects

3.1 INTRODUCTION

The design of electronic devices rests on an understanding of parasitic phenomena. The designer decides on current levels, stage gains, circuit geometry, attenuator values, capacitor sizes, etc., and all of these choices relate to parasitic effects. If the effects are included in a design analysis, a designer could perhaps conclude how to proceed. The procedure, however, is complex and requires an understanding of which parasitic effects are critical. Since the effects must be selected and estimated, we must immediately return to the basic understanding of the phenomena in order to make the estimate a good one. Needless to say, many a design goes through several prototype efforts before the parasitic effects are properly related to the design. It is the intent of this discussion to provide some background for the designer so that he is forearmed.

3.2 WIRING RESISTANCE

Many wire routing arrangements (printed-circuit tracks are included) can lead to trouble. There are a few rules to obey which will protect the designer. These involve the controlled flow of current to eliminate unwanted pickup or unwanted feedback processes. Examples illustrate these effects.

1. Unwanted signal pickup in input common leads. In Figure 3.1a, the input signal consists of all emfs sensed in the loop ① ② ③ ④. A filter electrolytic capacitor is shown connected to point ③ so that

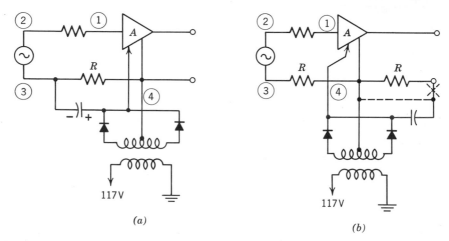

Figure 3.1   Parasitic resistance.

alternating current flows in R, the resistance of a printed-circuit track. To illustrate how low R must be, consider a 0.1% pickup for 10-mV signal levels. The current in such a circumstance might be 1 A. The permitted signal across R is 10 μV and the resistance should be below 10 $\mu\Omega$, an impossibility. The only solution is to return the capacitor to another point so that no signal results from current flow. Even if the electrolytic is returned to the output side (Figure 3.1*b*, the resistance must be below 10 mΩ, a figure which is difficult to attain. The proper solution is shown by the dotted connection in Figure 3.1*b*. The rule to be followed is simple: *Signal currents should only flow in signal conductors.* Power signals should flow in separate conductors if at all possible.

2. Signal current flowing in input common. In Figure 3.2 the output load current must flow in the path ① ② ③ ④ ⑤ ⑥ ①, and thus output signal is sensed at the input. The result is unwanted current feedback. For a positive-gain amplifier, the input signal is reduced by the flow of output current and this lowers the gain. For negative-gain amplifiers, the gain is increased. The transfer function is given by the expression

$$\frac{E_{\mathrm{O}}}{E_{\mathrm{IN}}} = \frac{R_L G}{R_L + R(G + 1)}$$

If $R = 0$, $E_{\mathrm{O}}/E_{\mathrm{IN}} = G$. If $R_L = \infty$, $E_{\mathrm{O}}/E_{\mathrm{IN}} = G$. If the gain is 1000 and the gain change is to be less than 0.1%, then $R = R_L/1000(G + 1)$.

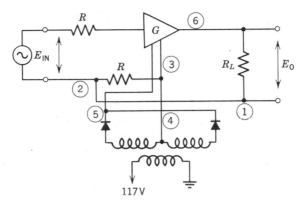

Figure 3.2   Load current flowing incorrectly in the input circuitry.

Typical values of R_L might be 1000 Ω. This implies that

$$R = \frac{1000}{10^6} = 10^{-3} \ \Omega \text{ or less}$$

It is interesting to see how this current feedback affects output impedance. Assuming the intrinsic output impedance without current feedback
is 0 Ω, the output impedance with feedback is given by the expression

$$Z = (G + 1)R$$

If $G = 1000$ and $R = 10^{-3}$, the ouput impedance is 1Ω. Thus it is obvious
that proper wiring is necessary to retain a low output Z. For negative-
gain amplifiers the output Z becomes negative; that is, when a load
resistor is placed on the output terminals, the signal increases.

The rule to follow for signal paths is simple. Do not allow input
and output signal currents to use common conductors, particularly when
large gain differentials are involved. One milliohm of common resistance
can alter performance significantly.

3.3   COMMON POWER SUPPLIES

Lead resistances to a common power supply in multichannel systems
provide a source of cross talk. The difficulty arises principally because
load currents in output stages must always return to the power supply.
These currents cause a potential drop along the connecting leads. The
potential drop is signal superposed on the supply voltages and is coupled
to other devices. These signals are picked up and amplified and are
obviously undesirable.

The most severe cross-talk test in multichannel systems is to drive all channels but one to full scale with fully loaded signals and observe the signals in the remaining channel. The test should be run at all frequencies of interest and at full gains to observe worst-case behavior. To eliminate cross talk, several design approaches are available.

1. Input stages can be separately powered or regulated to avoid pickup. This can be done for individual amplifiers or for all input stages in parallel. The former solution is more desirable.
2. Separate leads to the power supply can be used for each amplifier.
3. Separate power supplies can be used for each amplifier.

In general, large physical systems with common power supplies are undesirable. The cross-talk problems and ground-loop difficulties that result yield poor instrumentation. If the signals are ac coupled with transformers, the difficulties can be avoided.

Direct-current instrumentation with a common power supply is practical when differential amplifiers are used but only under controlled circumstances. The amplifier segments assigned to the common supply should have low gain. Also common output lead lengths should be kept minimum. This problem becomes more severe at high frequency as the line inductance affects the line impedance.

3.4 OUTPUT FEEDBACK SENSING

Consider Figure 3.3a in which the feedback connection point ① is at the output terminal proper. The output Z is given by the expression

$$Z_0 \simeq \frac{R_E}{G}\left(\frac{R_2}{R_1}\right) \tag{1}$$

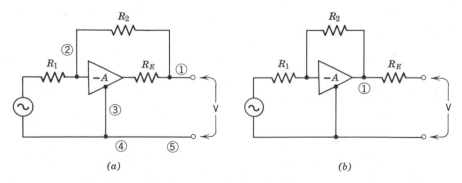

Figure 3.3 Output feedback sensing.

If $R_E = 1\ \Omega$, $G = -10^5$, $R_2/R_1 = 10^2$, then $Z_0 = 0.001\ \Omega$. In Figure 3.3b, the output Z is just R_E, or $1\ \Omega$. The correct wiring connects the feedback-sensing resistor to the point of measurement. In effect, the source resistance R_E is reduced by the feedback factor $A\beta$, in this case GR_2/R_1. Figure 3.3 could be a power supply as well as an amplifier. If the output noise signals are to be kept low, loop area ① ② ③ ④ ⑤ ① must be very small to eliminate the possibility of magnetic-flux capture either from internal or external sources. Any changing flux captured by this loop area becomes an output signal. By definition point ② is essentially at 0 V. Voltage ① and the equivalent flux capture voltage must be equal and opposite in sign to cancel so that point ② is nulled.

The geometry which will help eliminate parasitic-flux capture is to run lead ① ② and resistor R_2 next to or preferably coaxial with lead ④ ⑤. This treatment eliminates the pickup possibility as the loop area is minimum.

If unwanted pickup is sensed in the feedback loop, no amount of filtering or bypassing will cure the problem. The only solution involves a correction in circuit geometry. Signals sensed in the feedback loop appear in the output as if they were signals impressed on the normal input terminals.

3.5   CAPACITANCE

All signal processes involve parasitic capacitances. These capacitances may exist between transistor elements, or they may be input-to-output capacitances. If the current flowing in a stage of gain is increased by lowering circuit resistances, the effect of fixed parasitic capacitances can be shifted to a higher frequency. The proper approach in circuit design is to keep the parasitic capacitance minimal by proper circuit geometry.

Figure 3.4 shows a typical input stage with a capacitance C to a nearby signal point, gain G removed. The transfer function is

$$\frac{E_O}{E_{IN}} = \frac{GE_{IN}}{1 + GRCs} \qquad (2)$$

Where s is the complex frequency operator.

The RC time constant is increased by the gain factor G and becomes GRC. Thus if RC equals 10^{-4} sec and $G = 1000$, the new time constant is 0.1 sec. Thus gain multiplies the capacitance effects directly. Points that are separated by gain differences should always be separated physically or properly shielded.

The capacitance in Figure 3.4 might have been considered the base-to-

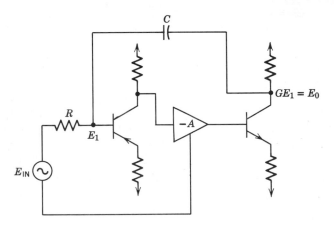

Figure 3.4 Gain effects on capacitance.

collector capacitance in the input transistor. The multiplication effect that results in this case is more commonly known as the "Miller effect." Obviously circuit geometry alone cannot reduce this effect. An improved transistor type or a lower circuit resistance could reduce the time constant to acceptable limits.

Shielding is often used to redefine the reactive paths.   The presence of a shield plane actually adds capacitance to the circuit, but if properly handled the capacitances that result are not increased by a large gain multiplication factor.

Figure 3.5 shows a ground plane added to the circuit of Figure 3.4.

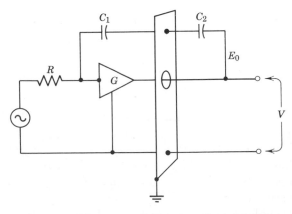

Figure 3.5 A capacitance guard plane (shield).

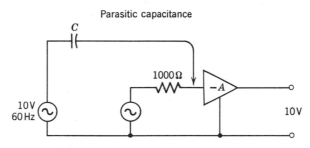

Figure 3.6 Parasitic input capacitance.

The two capacitances terminate on signal common and the gain multiplication effect is avoided. Consider a high-input-impedance amplifier operating from a 1000-Ω source with a gain of 1000. How much capacitance is permitted between the input and a 10-V 60-Hz source so that no more than 0.1% of full-scale signal will be present in the output of the amplifier? (see Figure 3.6). The full-scale input error signal is 10 mV. A 0.1% error is only 10 μV and represents a current flowing in the 1000-Ω source of 10 nA. The reactance of the permitted capacitance must be such that a 10-V source causes 10 nA to flow; the value of such a reactance is 1000 MΩ, a very high impedance level. At 60 Hz, 1000 MΩ is about 2.5 pF—which is a small capacitance! This capacitance is critical and note that the frequency and impedance level are moderate. The causative factors are the small signal levels and the 0.1% pickup permitted.

3.6 PARASITIC INDUCTANCE

Consider an amplifier with a 10-Ω feedback element when it is desirable to maintain a 1% flat frequency-amplitude response to 10 kHz. Since feedback-resistor values and frequency characteristics are interrelated, the 10-Ω resistor must not rise in impedance above 10.10 Ω at 10 kHz. This rise can be caused by an inductance at 10 kHz of just over 4 μH. If care is not taken, 4 μH can be picked up in as little as several inches of lead wire, not to mention the contribution of a wirewound resistor.

Capacitors are usually inductive at rf; thus they are ineffective unless special techniques are used. The engineer often feels the need to shunt an already low impedance point to reduce noise or rf content. The capacitor values are usually large and this adds to the difficulty.

Consider a Zener diode with a dynamic impedance of 1 Ω. To attenuate diode noise at 10 kHz effectively the reactance of a shunting capacitor should be 0.1 Ω. This is 160 μF, which obviously demands an electrolytic

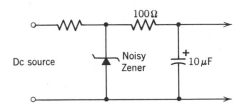

Figure 3.7 A practical Zener filter.

capacitor. Most electrolytics have a limiting impedance of about 0.5 Ω, so this is impractical.

If the frequency is higher, say 1MHz, the shunting capacitor now becomes 1.6 μF. A reactance of 0.1 Ω at 1 mHz is 0.016 μH, which is an extremely minute amount of inductance. Obviously capacitors with such low series inductance must be specially constructed.

If very low inductances are impractical, other means are required to reduce noise to acceptable limits. On example of how this might be done is given in Figure 3.7. Here a resistor is added so that a practical capacitor size can be used. Usually this added impedance causes no operating difficulty and regulation against source variations is still provided by the Zener action.

Some reduction in series inductance with metal-can electrolytic capacitors can be effected by a pseudo four-terminal treatment. Here the lines to be filtered are wired so that the currents must flow through the metal can of each capacitor. For low inductance several capacitors are paralleled to enhance the four-terminal effect. This treatment is shown in Figure 3.8.

The parallel connection of elements in several transistor or vacuum tubes requires parasitic suppression as a safety precaution. The preferred treatment is to add series resistors with paralleled elements so that no more than one element type is directly connected. The treatment for vacuum tubes and transistors is shown in Figure 3.9. Added cathode

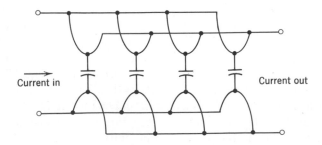

Figure 3.8 A four-terminal electrolytic filter.

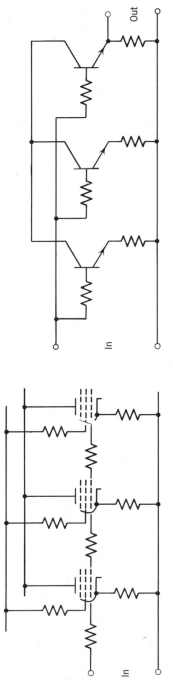

Figure 3.9 Added parasitic suppressors for vacuum tubes and transistors.

or emitter resistors constitute local feedback and should be small enough not to interfere with circuit performance. Grid or base resistors are typically 1 kΩ, screen resistors in tubes about 100 Ω, cathode or emitter resistors 10 Ω. Without these resistors oscillations at extremely high frequencies can take place, causing erratic and improper operation. It is very difficult to predict how the trouble may manifest itself if this precaution is not taken. It is also very difficult to observe the oscillations directly. The precaution is inexpensive and eliminates this source of difficulty.

3.7   POWER-SUPPLY DECOUPLING

Many amplifier circuits (linear integrated circuits included) have gain-bandwidth products in excess of 30 MHz. The closed-loop stability of such amplifiers can be dependent on the source impedance of the power supplies used to operate the amplifier. If long leads are used (in excess of 6 in.), the lead inductance can permit internal coupling that can cause the amplifiers to be unstable. The proper use of these devices involves a bypass capacitor from each power-supply line to signal common located at the amplifier terminals. A 0.01-μF ceramic capacitor is usually adequate. The complete circuit is shown in Figure 3.10. In multichannel systems in which common supplies are used the decoupling problem is complicated by lead and supply inductance. The comments of Section 3.2 are all applicable and the same design considerations apply. One added difficulty involves system stability. The cross-talk phenomenon can often be controlled, but the addition of a last channel or a change in load resistance can cause the entire system to oscillate.

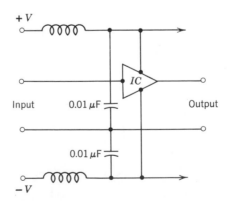

Figure 3.10 IC bypassing to prevent power-supply coupled instability.

The output impedance of a power supply at frequencies above 100 kHz is usually inductive because the gain in the electronics is falling and is not available to keep the output impedance low (see Section 4.15). The output sensing point for a group of channels is at best a compromise, and the leads to each channel appear as transmission lines. *RC* decoupling for each input stage can often remove any difficulty, but if this option was not included in the design, it may be difficult to implement after the fact.

3.8   INPUT PROBLEMS

High-frequency signals entering the input stages of an amplifier can introduce nonlinear behavior including dc offsets. A great deal can be done in the layouts to eliminate most of the pickup. It must be remembered that offending signals can be in the frequency range of interest and input filters will not handle this type of problem.

Most amplifiers have input stages that are required to handle only small signals. For this reason they have limited slewing-rate capabilities (see Section 8.17). If the volts-per-second requirements are exceeded (current is not available to drive parasitic capacitances), the stage will usually shift its operating point. Such a shift has the appearance of a signal, one not provided in the input terminals, and the resulting output is an error. If the input signal is high enough in frequency, the principal effect will be due to signal rectification at the input. This rectification and resulting shift in operating point can reduce gain, cause output shifts, or cause a generally noisy output signal to result.

Input stages may not be the entrance point for external disturbances. Any stage is vulnerable to pickup if the mechanics are present for sensing the external field, particularly if the unwanted signal is magnetic, as it can cut through an electrostatic shield with ease. In searching for the point of entrance from external fields *all* signal points should be considered.

Rf shielding can be ineffective if the input signal lines themselves are the carrier of the rf energy. Input *RC* decoupling is practical and is often used. A series resistor located at the input point together with any input capacitance can serve as the *RC* filter.

The signal geometry should be nearly coaxial (minimum loop area) to eliminate pickup of rf energy from sources other than the signal lines. This treatment should apply to all points along the signal path as well as to the feedback circuit. Differential signal treatment avoids many parasitic difficulties, but it is usually ineffective at high frequencies.

3.9   RF BYPASS DISCUSSION

The most significant comment that can be made about rf energy is that it is carried in the space between conductors and not in the conductors. The energy is conveyed by the distributed capacitances and inductances defined by the conductors. This is another way of saying the energy is carried by the electromagnetic field surrounding the conductors.

A bypass capacitor applied between two points simply forces a voltage node to exist between these two points. Rf energy can easily pass this point and appear in modified form elsewhere in the circuit. To terminate rf energy properly, the transmission complex carrying the energy must reflect all of the energy. A typical scheme for doing this is shown in Figure 3.11. Note that the reflection process is three-dimensional. The bypass capacitors used are feedthrough devices. RF energy is reflected at the metal wall used to mount the feedthrough capacitors. Low-frequency signals are easily passed through the metal wall.

Unwanted rf energy can be reflected by the presence of a line filter (e.g., an ac line filter). If the filter can be located in a manner similar to that in Figure 3.11, it can be effective. A metal wall can represent an electrostatic shield for signals. The same shield should not be used for both rf bypassing and signal shielding. The correct solution is to provide two shield systems, one for rf and one for signal shielding. Rf shielding can parallel as rack cabinetry, wire raceways, conduit, etc. In rf shielding, multiple ground ties to earth are permitted. In signal shielding, only one ground tie is permitted. Again, rf and signal shielding require separate treatment.

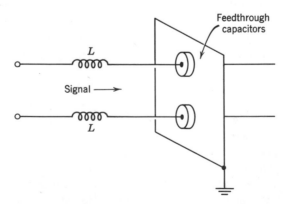

Figure 3.11   Proper rf filtering for energy reflection.

To obtain some measure of the component values involved, the characteristic impedances encountered in rf rarely go outside the range 50 to 1000 Ω. At 1 MHz, typical component values in Figure 3.11 might be 10 μH and 0.001 μF.

PROBLEMS

1. The full-scale current from an amplifier is 100 mA. If the signal is to change 0.1% of full scale from full load to no load, calculate the permitted source resistance for both a 10-V and a 1-V full-scale voltage.
2. The resistance of #20 wire is 10 Ω per 1000 ft. If the output resistance of an amplifier is to be 0.050 Ω, what is the maximum length of twisted wire permitted? (Assume the amplifier alone is 0.02 Ω and a connector adds 0.02 Ω.)
3. A power supply has a remote sensing feature. If the signal cable resistance is 10 Ω and the feedback factor is 100, what is the output resistance with feedback? With a 100 Ω load, what percentage additional output voltage appears at the power-supply terminals?
4. A 10-μF capacitor has 0.1-μH series inductance. What is the self-resonant frequency? What is the impedance of this circuit at twice the resonant frequency? At half the resonant frequency?
5. A Zener diode is rated 10 mA at 6.2 V, and 20 mA at 6.25 V. What is its dynamic resistance?

4

Feedback

4.1 INTRODUCTION

The engineer familiar with feedback techniques would state that the advantages are obvious. The uninitiated often has trouble understanding the process and finds feedback theory very complex. The first part of this chapter treats feedback ideas in their simplest form. Hopefully, the discussion will add insight into a very important phase of engineering.

Feedback and our daily routine are constant companions. When we drive a car between two white lines on a highway, we monitor the car by eye to keep the error distances in bounds. When we write on paper, we monitor the pencil pressure to keep a steady, even line. When we catch a baseball, we monitor the expected trajectory and move to intersect the ball in time. These processes each involve the human in a complex error-correcting system. Specifically, each system has a detection mechanism and the means to make the correction. Hopefully, the results are a safe trip, legible writing, or a fielded ball. This error-correction process is simply feedback, and it is fundamental to most of our daily activities.

An error-correcting system that is strictly electronic is called a feedback amplifier. An expected output signal is monitored for accuracy. If an error exists, a correction is made to reduce the error to acceptable limits.

4.2 THE DC AMPLIFIER

No time limit is imposed on the error-correcting problems just discussed. The driver of the car may go for hours, the writer may spend an evening writing, and the fly ball may be very high. Systems that perform for long periods of time must have no restrictions on their low-frequency response characteristics. It is necessary to use direct-coupled devices (dc amplifiers) to meet this condition. If an ac device is used, errors build up with time. Unless the problem is completely over in a well-defined interval, ac devices are not acceptable.

4.3   DC AMPLIFIERS AND FEEDBACK—INTRODUCTION

Direct-coupled amplifiers are in wide use today mostly in the form of *IC* amplifiers. To use these devices properly, an understanding of feedback is required.

To discuss the details of feedback in dc amplifiers, several areas causing frequent trouble must be considered. The first area of difficulty is the relation between signal levels and operating levels. In a typical stage of amplification, the operating point may be located anywhere from $+5$ to $+8$ V, but the largest signal swings may be only ± 0.5 V about the operating point. It is also possible to have a static operating point limited to a range of ± 0.5 V and signal swings of ± 10 V. It is important to remember that signals are just changes in operating points. For most signal discussions the operating points are unimportant.

A second area of difficulty is output offset or output zero. Feedback processes in themselves do not provide zero voltage output for zero-signal input. In direct coupling it is possible to arrange any static operating level for the output. A zero-output level for zero-input level is just one more constraint on the design.

A third area of difficulty is in the field of measurement. Consider an amplifier with a 1-y output offset. It has a gain of 10 and an input signal of 0.5 V is applied. The output shifts 10×0.5 or 5 V to a new value of 6.0 V. A measure of the 6 V would indicate that the gain is 12. The only valid gain measure subtracts the offset voltage to obtain the correct value 10. Mixing operating points with the measurements leads to trouble.

Measurements in other specification areas can be troublesome such as input impedance, output impedance, and dc gain. In the ideal amplifier troubles do not take place as all "good" things are true.   In practice the output has zero error and the input has an internal current source. Measurement technique is covered in Chapter 8.

4.4   DYNAMIC MEASUREMENT

Valid measures of performance in a dc amplifier involve differences. In ac measures static values are automatically ignored and only differ-

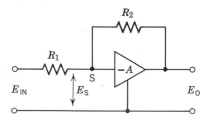

Figure 4.1 The basic operational feedback configuration.

ences are monitored. Consider the problem of measuring dc input resistance. The only valid procedure is to measure a ΔE input and observe the ΔI input that results. The ratio $\Delta E/\Delta I$ is the input impedance. Any method that does not subtract static input current will be in error. Consider the measure of output impedance. One valid procedure is to measure change in output voltage for a change in output current. The ratio $\Delta E/\Delta I$ is the output impedance. Any method that does not subtract zero offset errors will be in error.

4.5   OPERATIONAL FEEDBACK

The simplest operational feedback structure requires a high-gain inverting amplifier as shown in Figure 4.1. The resistors R_1 and R_2 are called operational feedback resistors. Consider as an example $R_1 = 1$ kΩ, $R_2 = 100$ kΩ, $A = 10^5$, and $E_0 = 1.0$ V. The voltage at S is simply $E_0/(-A)$ and in this example never exceeds 10 μV. The current in R_2 is known quite accurately since the potential at one end of R_2 is 1.0 V and at S 10 μV. If the 10 μV is neglected, the current in R_2 is $I = E_0/R_2 = 1.0$ V$/100$ $k\Omega = 10^{-5}$ A. This current has one legitimate source (assuming it did not arise from within the amplifier), and that is from the input node E_{IN} through R_1. For the current I the potential drop across R_1 is $IR_1 = (E_0/R_2)R_1 = -10^{-5} \times 10^3 = -0.01$ V. This is the input voltage except for 10 μV at the point S which was neglected. This exercise illustrates that the gain or ratio between the output and input voltages is very close to $-R_2/R_1$. For this example the gain is -100.

Four important results appear:

1. The gain is approximately the ratio between two resistor values and is not dependent on $-A$ (for large A).

2. The point S remains at near zero potential. If the input resistance of the gain block $-A$ is 10 kΩ, the current flowing in the resistance is

10 μV/10 kΩ = 10^{-9} A. This value is only 0.01% of the input current and further supports the statement that the same current flows in feedback resistor R_2 as in R_1.

3. Since the point S is essentially at null, the input current is simply E_{IN}/R_1 or the input resistance is just R_1.

4. The gain is negative; that is, the input signal appears inverted at the output.

Operational feedback can be characterized as follows: A known accurate current E_1/R_1 is taken from the source. (This assumes that the point S stays very nearly at 0 V.) If the circuit is that of Figure 4.1, only one input voltage can cause this current to flow. If the wrong current flows, the point S indicates an error and the amplifier responds by shifting the output voltage in a more promising direction.

Another viewpoint may be helpful. Consider a teeter-totter with a fulcrum point S as in Figure 4.2. When the end nearest the fulcrum is moved a distance E_{IN}, the far edge moves E_0. The ratio of E_0/E_{IN} is simply R_2/R_1 or the ratio of two lengths. This ratio requires that the point S be fixed. These mechanical displacements are an analog of the voltages in the electronic case.

4.6   SUMMING AMPLIFIERS

The operational feedback technique just described can be used with several inputs. Figure 4.3 shows three input signals applied to the same amplifier. The current flowing in R_2 is the sum of currents flowing into the point S from each of the three sources. These individual current values are known since the potential at S is near zero. If only one of the signal sources is active, the other two source resistances parallel the input resistance of the gain block $-A$. From earlier arguments, only very small signal currents flow in these resistances and the majority

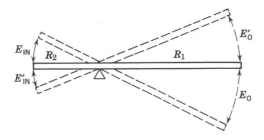

Figure 4.2   An operational feedback analog.

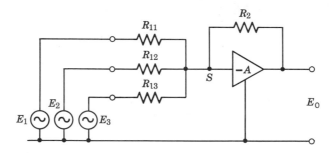

Figure 4.3 A multiple-input operational amplifier.

of the currents must therefore flow in R_2. This conclusion supports the idea that the signals do not react with each other and the output signal is the sum of three separate input signals. For this reason the point S is called a *summing point* and the operational feedback amplifier is also called a *summing amplifier*.

The point S moves only enough to provide an error signal for the amplifier. Any node point that remains nearly constant when influenced by other network driving points turns out to be a point of relatively low impedance. In the limiting case for infinite gain A the node point S is a zero impedance point or a short circuit. The assumption of a short circuit provides a convenient shortcut for writing down the operating equations for an operational feedback configuration. This will be apparent in the subsequent discussion.

4.7   POTENTIOMETRIC FEEDBACK

Figure 4.4 shows a potentiometric feedback structure. The gain block A amplifies the potential difference $E_1 - E_2$; that is, $E_0 = A(E_1 - E_2)$.

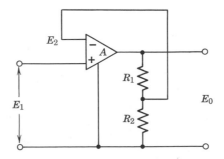

Figure 4.4 A potentiometric feedback structure.

If the gain A is very large, then E_2 very nearly equals E_1 for all values of E_0; for example, if $E_0 = 1.0$ V and $A = 10^5$, then $E_1 - E_2$ never exceeds 10 μV. The divider R_1 and R_2 can be shown to control the gain. By the preceding arguments, a potential very nearly equal to E_1 must appear at the node joining R_1 and R_2. If E_1 appears across R_2, the current flowing in R_2 by Ohm's law is $I = E_1/R_2$. This current flows in both R_1 and R_2, and the total voltage $E_0 = I(R_1 + R_2) = (E_1/R_2)/(R_1 + R_2)$. Overall gain is the ratio $E_0/E_{IN} = (R_1 + R_2)/R_2$. Thus for large A the gain is independent of A.

4.8   INPUT IMPEDANCE OF POTENTIOMETRIC DEVICES

The input circuitry which implements Figure 4.4 is usually differential in nature. A typical version might be that of Figure 4.5 in which Q_1 and Q_2 are matched transistors. The input signal E_1 and return signal E_2 are here very nearly equal. The difference between E_1 and E_2 is amplified by the differential stage Q_1 and Q_2 and by A and results in an output signal.[1] If the emitter of Q_1 were grounded, the dynamic base-emitter resistance (the $\Delta E/\Delta I$ of Q_1) would be the input resistance of the amplifier. Because the emitter follows the base, an input signal change

---

[1] The static potential differences and current flow between base and emitter are not of concern here. When the input base changes potential by 1 V, the emitter also changes potential by approximately the same voltage.

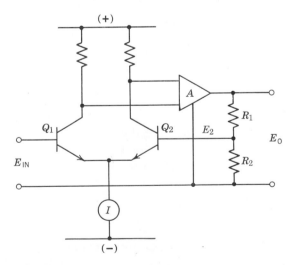

Figure 4.5   A potentiometric amplifier with input stages shown in detail.

produces a reduced signal current change in the base-emitter resistance. In effect, the input resistance is raised by the presence of potentiometric feedback.

Some typical operating values might clarify this point. Assume the dynamic base-emitter resistance in a grounded emitter configuration is 1 kΩ. In the feedback circuit the emitter is returned to within 1% of the input value by the amplifier. For a 1-V input signal change, the emitter moves to 0.99 V. The change of current flowing in the base-emitter resistance is 0.01 V/1 kΩ or 10 μA. As viewed from the source, an input signal change of 1.0 V caused a difference current to flow of only 10 μA. By Ohm's law this is an input impedance of $1.0/10^{-5} = 100$kΩ.

Potentiometric feedback is characterized by a matching process. In Figure 4.5, when E_2 matches E_{IN}, the output is a measure of E_{IN}. Some method must always be provided to sense differences so that an error measurement can be made for amplification. However, the sensing has its own inherent errors, which limit the performance of the potentiometric system.

4.9 POTENTIOMETRIC MECHANICAL ANALOG

Consider a marble on a long board. If the board is held level, the marble will not roll in either direction. If one end is raised an amount Δe, the marble will accelerate downhill until the other end of the board is moved a matching amount Δe. All the potentiometric processes are present, that is, an input signal and a "matching" output signal. The ball acceleration is an error signal and this is amplified (the board is moved) until the marble is stopped or recentered.

If the board is irregular, then a level spot for the marble is not a measure of level for the entire board. This is an example of performance being limited by the quality or match in the sensing element. It is obvious that the only improvement possible is to smooth the board. The parallel to this error in electronics involves the matching of input differential transistors such as those in Figure 4.5. If this match is imperfect, it directly affects the input-output signal match.

A few features of potentiometric feedback have been presented: (*a*) the gain of a potentiometric system can be made the ratio of resistors; (*b*) input impedance enhancement usually results; (*c*) the quality of performance is limited by the matching errors in the input sensing circuitry. More will be said about potentiometric feedback in later sections.

4.10 FEEDBACK—DEFINITIONS

In the preceding discussions the ideal characteristics of potentiometric and operational feedback were considered. If the amplifier gain without feedback was very high, then the qualities of the circuit were determined by the feedback circuitry. In practical situations the amplifier gain does influence the performance in many ways.

For feedback to be effective the gain before feedback must be large compared with the gain determined by the feedback elements. As a rule of thumb, if the gain excess is a factor of 100, the gain after feedback will be accurate to within 1%. If the excess gain is 1000, the gain after feedback will be accurate to 0.1%.

To describe the performance of feedback structures, the following definitions will help. These definitions are not intended to be precise, but to convey understanding to the reader.

Definition. Open-Loop Gain.

The gain of an amplifier before feedback is introduced is open-loop gain. Open-loop voltage gains typically exceed 10^4 and frequently exceed 10^7. When the gain is represented by a current-to-voltage ratio, gain is expressed in units of transconductance or ohms.

Definition. Closed-Loop Gain.

The gain of an amplifier with feedback elements connected is closed-loop gain. Typically closed-loop gains are less than open-loop gains by a factor greater than 100.

Definition. Loop Gain or Feedback Factor.

This is essentially the ratio between open-loop gain and closed-loop gain. It is that excess of gain available to be "fed back" to force the amplifier to "behave" as the feedback components dictate.

Definition. Feedback Fraction or β.

This is the fraction of the output signal reaching the input terminals. It is usually formed by a resistive attenuator. Figure 4.6 shows a feedback amplifier $-A$ with a feedback structure β. The input signal E_B to the amplifier is the sum of input signal E_{IN} and a fraction β of the output signal or βE_0. Thus

$$E_B = E_{IN} + \beta E_0 \tag{1}$$

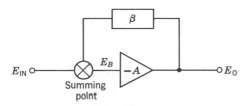

Figure 4.6 A summing-point feedback representation.

The amplifier requires that

$$E_O = AE_B$$

By substituting (2) into (1) the gain ratio becomes

$$\frac{E_O}{E_{IN}} = \frac{-A}{1 + A\beta} \tag{3}$$

This equation has all the elements of the preceding definitions.

1. $-A$ is the open-loop gain.
2. E_O/E_{IN} is the closed-loop gain.
3. $A\beta$ is the loop gain or feedback factor.
4. β is the feedback fraction.

Equation 3 may be rewritten by dividing both numerator and denominator by A. The closed-loop gain is

$$\frac{E_O}{E_{IN}} = -\frac{1}{1/A + \beta} \tag{4}$$

It is apparent that when A is large, $1/A$ is very small. When $1/A$ is small compared with β, it can be neglected, and we are left with

$$\frac{E_O}{E_{IN}} \cong -\frac{1}{\beta} \tag{5}$$

Thus the closed-loop gain is independent of A for A large. From (4), if $1/A$ is 1% of β, then the closed-loop gain is low by 1%. If reciprocals are considered, when A is 100 times $1/\beta$, the gain will be low by 1%. When A is 1000 times $1/\beta$, the gain will be low by 0.1%. Thus the ratio of A to $1/\beta$ is the key to determining the gain accuracy of a feedback system. This ratio is

$$\frac{A}{1/\beta} = A\beta \tag{6}$$

and, as indicated before, is termed the loop gain. This term is descriptive, since $A\beta$ is the product of forward and reverse transmission in the loop shown in Figure 4.6.

4.11   DECIBELS

Decibels are strictly speaking a logarithmic measure of power ratios. Classically the definition is

$$dB = 10 \log_{10} \frac{P_1}{P_2} \tag{7}$$

where P_1 and P_2 are two power levels.

When this ratio is applied to a *common impedance level* (voltages across the same resistor), the ratio can be expressed in terms of these voltages or

$$dB = 20 \log \frac{E_1}{E_2} \tag{8}$$

It is common practice in discussing amplifiers and feedback to use (8) as a voltage ratio, neglecting the common impedance level requirement. When this is done, the dB statement is *not* a measure of power gain. It is just 20 times the logarithm of the voltage ratio. It is useful to use a logarithmic voltage ratio concept, the only difficulty being that the title decibel is not correct. However, until a new name is universally accepted, the word decibel will have to do.

The standard practice in amplifier specifications is to discuss gains in decibels. Typical figures might be

$A = -100,000$ or   90 dB of negative open-loop gain
$\beta = \frac{1}{1000}$ or -60 dB
$A\beta = 100$ or   40 dB of loop gain or feedback

$\dfrac{1}{\beta} = 1000$ or   60 dB of closed-loop gain

The following table shows typical, frequently used decibel values and their corresponding voltage ratios.

Ratio	Decibels
0.001	-60
0.01	-40
0.1	-20
0.301	-10
0.5	-6
0.707	-3
0.90	-1
0.99	-0.1
1.0	0
1.01	$+0.1$
1.1	$+1$
1.414	$+3$
2	$+6$
3	$+10$
10	$+20$
30	$+30$
100	$+40$
1,000	$+60$
10,000	$+80$

4.12   THE PRACTICAL PROBLEM

Figure 4.6 serves well as a representation but it is unfortunately an oversimplification. The summing point of a practical operational amplifier does not exactly sum an input and an output signal. Similarly, the feedback fraction is not easily found by inspection because the starting definitions are not exact. It is of interest to examine the operational feedback circuit (Figure 4.1) in more detail. Two equations describe the circuit performance:

$$E_O = -AE_s \tag{9}$$

and

$$E_s = E_{IN} - \frac{(E_{IN} - E_O)R_1}{R_1 + R_2} \tag{10}$$

Substituting (9) into (10) yields the transfer voltage ratio

$$\frac{E_O}{E_{IN}} = -\left(\frac{R_1 + R_2}{R_2 A} + \frac{R_1}{R_2}\right) - 1 \tag{11}$$

As A gets large, the transfer ratio approaches a gain of $-R_2/R_1$ as it should.

The β that would result from the definition above would be $R_1/(R_1 + R_2)$ and this is incorrect.

The summing point representation of Figure 4.6 should be used with caution. It is correct to assume that for large gains, the summing point signal approaches zero. This assumption will then allow an analysis of input and output signal ratios. The difficulty in Figure 4.1 is not hard to find. When the amplifier has a gain of zero, the input still couples to the output via the feedback path. Such a direct transmission path is not found in Figure 4.6. When the effects of direct transmission are included, then the difficulty disappears. Direct transmission must be considered in the general case, but for most practical understanding it need not be added to the discussion. It is a necessary consideration if the definitions are to be exact.

The summing-point signal in Figure 4.7 is the output signal divided by $-A$ or $E_O/-A$. Since $E_O/E_{IN} \cong -1/\beta$, the signal at the summing point is

$$\frac{E_{IN}}{A\beta} \tag{12}$$

In words, the summing-point signal is the input signal reduced by the

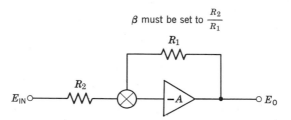

Figure 4.7   The correct β value for an operational amplifier.

loop gain. As the loop gain increases, the summing-point signal reduces. This calculation is correct for both feedback representations as long as the β value used is R_2/R_1 and the loop gain is high.

4.13   ERROR-CORRECTION PROCESSES

The simplest diagram of a feedback amplifier contains only an input and output terminal pair. Idealistically the only signals seen by the amplifier are supplied by the input terminals. In practice all other internal circuit points must be considered as possible input terminals. They include power-supply connections, and offset points or stabilization terminals. Signals appearing on these nodes may also appear in the amplifier output.

In a normal design one particular terminal pair is designated as "input." Other terminal pairs can also serve as inputs, but the performance from these points is generally favorable. If the designer is aware that all terminals are possible inputs, he can guard against unwanted interference. The gain from any internal points to the output can be calculated, and obviously some points can contaminate the output more easily than others. Any two points might be considered output terminals, but, here again, certain terminals provide the proper performance characteristics.

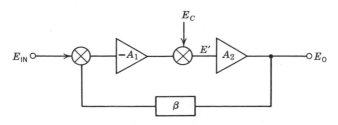

Figure 4.8   An idealized internal summing point.

To gain insight into the interference problem, an idealized internal summing point is included as in Figure 4.8. The gain from the normal input terminal to the output is

$$\frac{E_O}{E_{IN}} = -\left(\frac{1}{A_1A_2} + \beta\right)^{-1} \simeq -\frac{1}{\beta} \tag{13}$$

The gain E_O/E_C from the internal point can be approximated as follows: Assume the input signal to amplifier A_2 is E'. Then E' is the sum of E_C and the output signal going around the loop through the β path via the gain block $-A_1$. Stated mathematically

$$E' = E_C + E_O\beta(-A_1) \tag{14}$$

Since

$$E_O = E'A_2 \tag{15}$$

when (15) is substituted into (14),

$$\frac{E_O}{E_C} = -\frac{1}{A_1}\left(\frac{1}{A_1A_2} + \beta\right)^{-1} \simeq -\frac{1}{A_1}\frac{1}{\beta} \tag{16}$$

Thus the gain from an internal point to the output is the normal gain $1/\beta$ reduced by the gain preceding the point of injection.

The idealized summing process shown in Figure 4.8 is again misleading. Actual injection processes are not as simple as this although the principles illustrated still hold. Contamination usually occurs through a summing impedance in an operational sense. This circuit is shown in Figure 4.9. The signal at S can be assumed to be near zero if A_2 is large. Using this fact, we see that the current supplied by E_C equals the current supplied via the output loop through β and $-A_1$. Each current is

$$I = \frac{E_C}{R_2} = E_O\frac{\beta(-A_1)}{R_1} \tag{17}$$

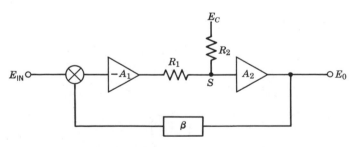

Figure 4.9 A practical internal summing point.

The ratio E_O/E_C now becomes

$$\frac{E_O}{E_C} = -\frac{R_1}{R_2}\frac{1}{A_1}\frac{1}{\beta} \tag{18}$$

This is the same form as (16) except for the factor R_1/R_2. When $A_1 = 1$, (18) reduces to the gain of an operational amplifier. The important feature to note here is this: A contaminating signal introduced at the input (for $A_1 \simeq 1$) is amplified as a normal signal with gain $1/\beta$. The amplifier cannot differentiate between "good" and "bad" signals when they are injected at or near the input.

Figures 4.8 and 4.9 are operational feedback circuits. The process of signal injection for potentiometer feedback is the same. The contaminating signal cannot be differentiated from a desired signal when it is injected at or near the input.

Power-supply ripple voltage is a common source of signal injection in feedback amplifiers. A power supply is usually connected in parallel to many points, but the effect is usually related to its influence on those stages nearest the input. In an *IC* amplifier the stages are not separable by the user, and the coupling from the power connections is a prime concern to the *IC* designer. The specification of power-supply sensitivity is usually given in microvolts per volt. If the specification is 50 μV/V, then, when the power supply changes 1 V, the output responds as if there had been an input signal of 50 μV. In most cases, if the supply had 1 V of 120-Hz ripple, the output would contain this same ripple signal multiplied by the gain fraction used.

In discrete component designs it is impractical to filter the power supply equally for high-current output stages and low-level input stages. It is standard practice to leave ripple voltages on the supplies connected to output stages and regulate or filter the supplies only when they involve the input stages. If there is sufficient open-loop gain, the power-supply signal will not appear in the output.

4.14 STABILITY IN FEEDBACK AMPLIFIERS

All of the preceding discussions excluded frequency-sensitive (reactive) elements. Since these elements exist parasitically, their effect must be considered. Procedures to stabilize a feedback amplifier require the addition of reactive components. Once the mechanism of instability is appreciated, the design procedure becomes obvious.

In the general equation

$$\frac{E_O}{E_{IN}} = -\frac{A}{1 + A\beta} \tag{19}$$

The gain A must be considered complex. To be more specific, the gain A is a function of frequency and therefore has phase shift associated with it. If a signal frequency can be found at which A has a phase shift of 180° and the product $A\beta$ is greater than unity, the amplifier will be unstable. Note that the amplifier gain is negative in Figure 4.7 and in this discussion inversion is *not* considered as a phase shift of 180°. The phase relationships at low frequency, assuming a dc amplifier, must be used as the 0° reference. A 180° phase shift for stability discussions must represent a *time delay* of $\frac{1}{2}$ cycle through the amplifier. Obviously the negative gain is present at all low frequencies and does not represent a delay.

At a high frequency each parasitic capacitance in an amplifier can contribute 90° of phase shift. Two such parasitic elements are sufficient to produce 180° of phase shift (time delay). If nothing is done to control the total phase shift, the feedback amplifier will be inherently unstable.

The presence of one parasitic capacitance across a gain-sensitive element will reduce the forward gain at a rate of 6 dB/octave. The phase shift in the region of attenuation approaches 90°. It is correct to associate attenuation rate with phase shift on a proportional basis. An attenuation rate of 12 dB/octave thus corresponds to a limiting phase shift of 180°.

Stabilizing a feedback amplifier consists of managing the gain so that the phase shift for $A\beta > 1$ never exceeds 180°. The phase margin requirements vary among designs. If the amplifier is to operate over a wide temperature range or with varying feedback parameters, the margin requirements are more severe.

Figure 4.10 shows a typical open-loop gain and open-loop phase response. This amplifier would be unstable for a closed-loop gain of 20 dB but stable for a closed-loop of 40 dB. This point is illustrated as follows: a closed-loop gain of 20 dB requires that $1/\beta = 20$ dB or $\beta = -20$ dB. The product $A\beta$ is unity (0 dB) when A has a gain of $+20$ dB. An examination of the A curve at the $+20$-dB point shows a slope of 12 dB/octave and a phase shift of 180°. This amplifier gain would be unstable. At a closed-loop gain of 40 dB, $1/\beta = -40$ dB. For $A\beta = 0$ dB, $A = +40$ dB. At this point on the A curve, the slope is 6 dB/octave and this amplifier gain would be stable.

To manage the gain so that the amplifier is stable at $1/\beta = 20$ dB, the gain A must be reduced at lower frequencies so that on the modified response the phase shift at $A = 20$ dB is less than 180°. The specific method of modifying the gain varies, but it often consists of a series resistor-capacitor combination, paralleled across a gain-determining element. This method reduces the gain on a controlled frequency basis.

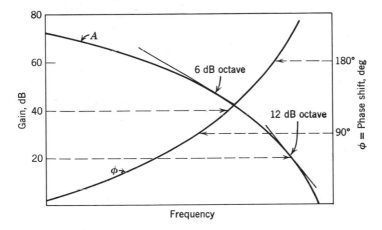

Figure 4.10   Gain and phase plot of a typical dc amplifier.

The choice of elements and their location are important details of each amplifier design. A managed gain curve is shown in Figure 4.11. This response would also indicate that the amplifier is stable for a closed-loop gain of 100 ($1/\beta = 40$ dB) but not stable for a closed-loop gain of unity.

4.15   A TYPICAL SHAPING NETWORK

The gain of a transistor stage is dependent on the impedance of the collector circuit. If the collector resistor is shunted by a parallel RC

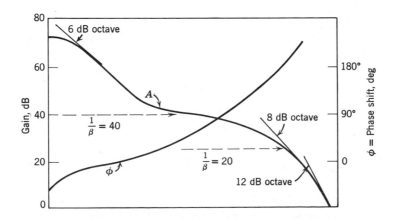

Figure 4.11   A modified gain and phase plot.

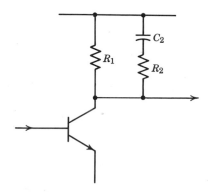

Figure 4.12 A simple gain-shaping network.

circuit, the gain at high frequencies is reduced. The reduction factor is just the ratio of low-frequency to high-frequency resistance. An example is shown in Figure 4.12.

If $R_1 = 9$ kΩ and $R_2 = 1$ kΩ, the gain is proportional at low frequencies to 9 kΩ and at high frequencies to the parallel resistance (900 Ω) or a 10:1 gain reduction. Since the gain attenuation rate is determined by one capacitor, the associated open-loop phase shift from this network never reaches 90°.

The frequency at which the gain reduction starts is determined by the $R_1 C_2$ time constant. At the frequency $f_1 = 1/2\pi R_1 C_2$ the gain is reduced 3 dB. The terminal frequency, when the gain is within 3 dB of final value, is $f_2 = 1/2\pi R_2 C_2$. In Figure 4.11 the gain shaping is shown starting too low in frequency. It is advisable in general to keep the open-loop gain as high as possible out to the highest frequency. This procedure increases the feedback factor over a wider frequency range. For the same reason it is desirable to keep the gain loss to a minimum consistent with sufficient phase margin. Two rules are suggested to aid in selecting a shaping network that meets this requirement:

1. Determine the minimum gain reduction requirement (use a large capacitor and determine the largest series R_2 that must be used to stabilize the amplifier).

2. Determine the highest frequency for the start of amplitude shaping (with this largest R_2 select a smallest capacitor value C_2).

An observation of amplifier performance is necessary while the above adjustments are made. The techniques of observation are discussed in Chapter 8.

4.16   AMPLITUDE-VERSUS-FREQUENCY RESPONSE

An open-loop amplifier usually has a very poor amplitude-versus-frequency response. The "rounded-off" response which results from cascading several blocks of gain is the result of the parasitic capacitance in each section. The step-voltage response usually has *no* overshoot and gradually reaches final value as the capacitances are charged to final value. For a typical single such RC the time to achieve 99% of final value is four time constants. If $C = 10$ pF and $R = 100$ kΩ, then $RC = 1$ μsec. The charging time to 99% of final value is 4 μsec. If three such circuits are involved, each circuit must recover to 0.3% of final value and this requires five time constants.

In amplifiers many capacitances are effectively multiplied by gain factors; for example, base-to-collector capacitances are multiplied by base-to-collector gains. A 2-pF transistor specification thus can easily affect the circuit performance as though 100 pF were paralleled across the collector resistor. In the preceding example the five time constants now becomes 50 μsec, a tenfold increase.

The closed-loop amplitude response of a feedback amplifier is not that of simple capacitances. As indicated in Section 4.14, if a group of parasitic capacitances is present such that $A\beta$ reaches 180° of phase shift for $|A\beta|$ greater than unity, the amplifier is unstable. An unstable amplifier is an oscillator, not describable as a cascaded RC network. As feedback is gradually increased around a stable amplifier ($A\beta$ is

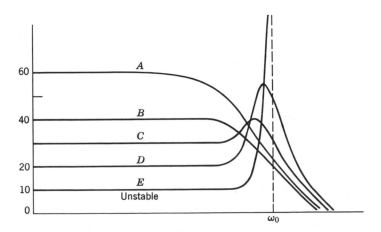

Figure 4.13   The amplitude response of an uncompensated feedback amplifier as a function of closed-loop gain.

increased), the bandwidth increases. As more feedback is applied, the amplitude response might exhibit a marked resonance at one frequency. As $A\beta$ is increased further, this resonance increases until the amplifier can be made to oscillate at this frequency. This phenomenon is shown in Figure 4.13. A response such as C is acceptable, but responses D and E are generally considered undesirable.

4.17   TRANSIENT RESPONSE

The step-voltage response for the various gains shown in Figure 4.13 is shown in Figure 4.14. Note that the overshoot is a sensitive measure of a peak in the amplitude response. The curves have all been gain normalized so that they correspond to typical observations on an oscilloscope. Curve C is acceptable, whereas curve D is not.

4.18   SQUARE-WAVE RESPONSE OF DC AMPLIFIERS

The step-voltage response ~~of Figure 4.14~~ can be obtained with a square-wave voltage generator. The repetitive nature of the square-wave source is ideal for synchronization with an oscilloscope sweep. If the repetition rate is low enough, then the response resulting from each transition is the same as the step response. Since the square-wave signal steps in both directions, adjacent patterns are inverted. ~~The response~~ in

T/f 's

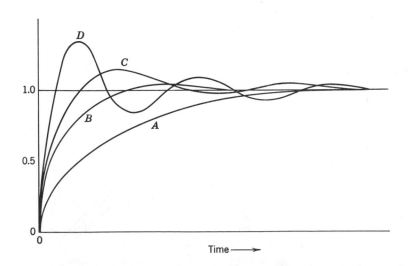

Figure 4.14   Normalized step response for the gains in Figure 4.13.

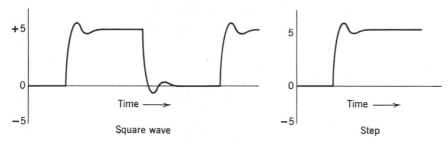

Figure 4.15   A typical transient response to square waves and to a step function.

Figure 4.15 is typical of many amplifiers. When the square-wave repetition rate is varied, the leading-edge information remains invariant. If the repetition rate is too high, the transients that are excited will have insufficient time to decay and the response will not clearly show the step response of the amplifier. The rule to follow is simple: Use a repetition rate that permits observation of the entire transient decay after each signal transition. If the response is different in opposite signal directions, the amplifier is nonlinear. This problem is discussed in Section 8.17. The magnitude of the excitation should be reduced so that the response is symmetrical in both directions.

If the response is symmetrical but the response varies with amplitude, the amplifier is again nonlinear, as shown in Figure 8.20. To observe this phenomenon, view the transient response, say, at a 1-V level. Increase the signal level to 10 V, but decrease the oscilloscope sensitivity to one-tenth its previous value. Under these conditions the signal should be *exactly* the same in waveform.

When an amplifier is linear, all gain elements are operating within their normal range. In nonlinear operation a gain stage may be current or voltage limited. These conditions are often described as a "slew-rate difficulty." In a linear device, 90% of final value occurs in the same lapsed time, independent of signal level. If a 1-V step signal reaches 0.9 V in 10 μsec, a 10-V step signal reaches 9 V in the same 10 μsec. If it takes any longer, the amplifier is nonlinear and it is slew-rate limited.

This distinction is a very important one. If an amplifier is operating in a nonlinear manner, its open-loop gain is reduced. With the gain reduced, the transient qualities are changed and an observation under these conditions can be very misleading. Since square-wave observations are intended primarily as a measure of stability, the measure is obscured.

Nonlinear performance is also very important and is not to be neglected. Before these results can be interpreted, the linear nature of square-wave testing must be understood. Recovery from overload, stability superposed on offset, etc., are all important parameters, and one should be able to distinguish recovery processes from normal linear phenomena.

4.19   RESPONSE OF AC AMPLIFIERS

An ac amplifier usually has zero gain at zero frequency. It has at least one coupling capacitor or it has a coupling transformer. Such coupling elements can store significant energy, and release of this energy can occur over long periods of time. For this reason the response after transient excitation takes a long period of time to decay to zero.

A simple RC coupling network illustrates this point. When a step function is applied to the RC network of Figure 4.16, the response decays to 1% of final value in approximately four time constants. If the sag in 10 msec is 1%, then $RC = 1$ sec. The time to sag to 1% of final value is thus 4 sec, a long time.

When a reactive element is included in a feedback amplifier, the feedback lengthens the recovery process. An open-loop low-frequency —3 dB response of 10 Hz is reduced to 1 Hz with 20 dB of feedback. This increased low-frequency response has a corresponding transient response time that is 10 times longer.

If several reactive coupling elements are included in an open-loop response, the possibility of instability with feedback is present. The

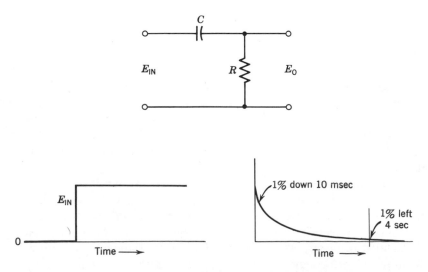

Figure 4.16   Response of a coupling network to a step-function input.

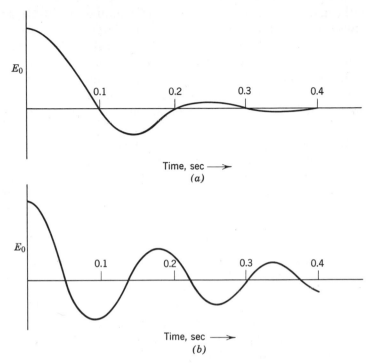

Figure 4.17   Low-frequency step-response character in ac amplifiers.

same rules apply to this problem as apply to the high-frequency case. If a frequency can be found at which $|A\beta| > 1$ and there is 180° of phase shift, the amplifier will be unstable. In effect, only one reactive element at a time can shape the open-loop gain for frequencies at which $|A\beta| > 1$. If two RC time constants are involved, they must be staggered such that the phase-shift rule is maintained.

Instability in ac amplifiers does not appear on the leading edge of a step function as is the case for high-frequency stability. Any tendency toward instability appears only after long periods of time have elapsed. The clue to instability involves the number of observable zero crossings after a step function has been placed into the amplifier. Figure 4.17a shows the step response of a stable ac amplifier; compare it with Figure 4.16b, a marginal case. These responses reflect a 10-Hz lower frequency response.

An unstable response has a sinusoidal buildup that increases until some limit point is reached. The limit point can be a power-supply voltage, a current limit, or a diode clamp.

When the gain is controlled so that the amplifier is just unstable, the amplifier becomes an oscillator. When this marginal stability is controlled by a nonlinear element that reduces the value of $|A\beta|$ to unity at just $180°$ of phase shift for one particular signal amplitude, the amplifier is called a signal generator. To vary the oscillator frequency the reactive elements causing the phase shift are varied. If all controlling capacitors are reduced by a factor of 2, the frequency of oscillation doubles.

Ac amplifiers with feedback must be designed for high-frequency and low-frequency phase margin. This requirement complicates the design by restricting the use of reactive elements. Coupling capacitors, emitter bypasses, and power-supply filter capacitors must all be considered as each can introduce phase shift into the gain term. It is good practice to make all but one time constant long. This one controlling time constant determines A such that for $|A\beta| > 1, \phi < 180°$.

When a transformer is introduced into the forward-gain path, both low-frequency and high-frequency elements are included. The low-frequency phase character is a function of the shunt magnetizing inductance, and the high-frequency phase character is a function of both the series leakage inductance and the primary and secondary coil capacitances. In some cases, primary-to-secondary coil capacitances cannot be considered as lumped into the shunt coil capacitances and must be included separately in the circuit analysis.

The serious problem with transformers in the forward gain arises from the nonlinear nature of the magnetizing inductance in the transformer. If the flux level reaches its limiting value, the magnetizing inductance disappears and that segment of the circuit is in effect short-circuited. Circuit performance as well as stability in this mode are difficult to describe and analyze at best. The designer should consider the possible operating modes in which the transformer core can be saturated and then test for these conditions. If the saturation causes no undesirable effects, the design is acceptable.

Coupling capacitors can be used to eliminate direct current from the coils of the transformer. This coupling element is another source of phase shift and therefore must be either large enough *not* to dominate the phase shift or small enough to be the dominant phase-determining element. The latter is preferable since such an element can attenuate low-frequency signals so that they do not saturate the transformer core. A large coupling element requires that the magnetizing inductance be large, which usually results in a larger core.

Often ac amplifiers are designed with an output coupling element. The feedback for the device is taken from a point inside this coupling

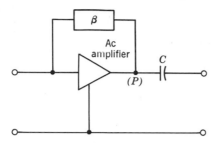

Figure 4.18   External ac coupling.

element, an arrangement shown in Figure 4.18. The advantages are two-fold. First, the element is not involved in the amplifier stability process, and, second, no dc offset appears on the output. The disadvantages are that the output impedance is not controlled by the feedback and the dynamic behavior of the amplifier is filtered by the capacitor. To over-come the latter effect, all stability observations should be made at the point of feedback (P), not at the output. It is possible that C could be small enough to obscure a marginally stable condition if the observa-tion were made at the output.

4.20   SIGNAL LEVELS INSIDE A FEEDBACK LOOP

In the open-loop case, the signal level is defined by the gain in each stage. If a stage has a frequency-dependent element, this quality is carried through to the output on a direct basis.

If there is an adequate loop gain, that is, $A\beta << 1$, the effect of a frequency-dependent element is not the same as that above. To avoid a mathematical description, a simple gain change can illustrate the effect.

Consider an amplifier with $1/\beta = 100$ and $A\beta = 100$. The gain of this circuit is

$$G = -\frac{A}{1 + A\beta} = -\frac{10,000}{1 + 100} = -\frac{10,000}{101} \simeq -99.01$$

If A is changed to 50, the new gain is

$$G = \frac{5000}{1 + 50} = -98.04$$

In this example a change of a factor of 2 in open-loop gain affects the closed-loop gain by 1%. Now consider that the forward amplifier is made up of two blocks of gain (see Figure 4.19). If the gain change

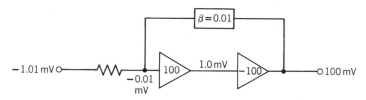

Figure 4.19 Signal levels for $-A\beta = 100$, $\beta = 0.01$.

occurs in the second block, the signal level before the second block must double. This is apparent from the signal levels shown in Figure 4.19. Feedback has this effect. Signal levels adjust *before* any point of change—they do not significantly alter after a point of change. This unchanged signal level is the only one that can cause a correct signal level to appear at the output. (The input signals in Figure 4.19 and 4.20 have been changed slightly so that the internal points can have integer voltage values.)

When the gain change is frequency-dependent, the signals ahead of the point of change all adjust accordingly. Because the gain is complex and there is phase shift, the signal ahead of the frequency-dependent element must have a nearly equal and opposite phase angle. This requirement occurs because the closed-loop output signal has very little resultant phase shift. The closed-loop phase shift at any frequency can be calculated from the open-loop complex gain; for example, if $A = 10,000$ at a phase angle of $30°$ and $1/\beta = 100$, then

$$G = -\frac{10,000 \underline{/30°}}{1 + 100 \underline{/30°}}$$

To calculate the phase angle for G, the denominator must be added vectorially. The denominator term is $100.5 \underline{/29.9°}$ and G becomes $99.52 \angle 0.3°$. Thus the internal phase shift is reduced by the feedback factor. The signals within a feedback loop with phase shift are shown in

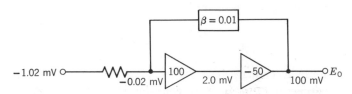

Figure 4.20 Signal levels for $-A\beta = 50$, $\beta = 0.01$.

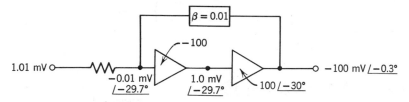

Figure 4.21   Signals within a feedback loop with phase shift.

Figure 4.21. Again the gain block with change is shown as the second block.

The extension of this idea to nonlinear elements within the feedback loop is now clear. A nonlinear element can be considered a gain change that is dependent on signal level. If an output stage is nonlinear, the effective gain error is reduced by the feedback factor; for example, a 10% effect is reduced to 0.1% if there is 40 dB of feedback; that is $A\beta = 100$. These facts point up how important it is to view the signal levels within the feedback loop in troubleshooting correctly, since they provide the only information available. Figure 4.22 shows the input/output voltage relationships of a typical nonlinear gain block. With adequate feedback the overall transfer ratio is nearly a straight line. The signal ahead of the nonlinear block exactly compensates for the distortion by being "oppositely" distorted. Since the points preceding this block are linear, they too *will contain* distorted signals. The only points in the feedback loop that "look like" the input follow after the point of distortion. The nonlinear block of Figure 4.22 is shown in a feedback loop in Figure 4.23 together with the associated signal patterns.

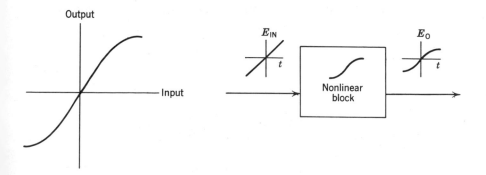

Figure 4.22   An open-loop nonlinear gain block.

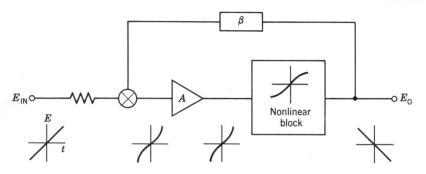

Figure 4.23   Waveforms associated with a nonlinear block in a feedback loop.

4.21 ELIMINATION OF EXTRANEOUS PICKUP

The examples above show how distortion in the forward gain is reduced by the feedback factor. When extraneous signals are introduced, the feedback processes tend to cancel these unwanted signals. The cancellation varies depending on the point of injection, which is discussed next.

Consider a square-wave input to a feedback amplifier. Assume the output stage has 120-Hz ripple on its power supply. The 120-Hz ripple would appear in the output added to the signal if there were no feedback. With feedback, the signals ahead of the output stage carry the proper "antisignal" to cancel the 120 Hz. The signal patterns are shown in Figure 4.24. When undesirable signals are added to the signal processes, their effect is related to the point of injection. When undesirable signals are added at the input, they cannot be differentiated from useful signals, and in effect they are not canceled. When they are added at the output, as in Figure 4.24, they are reduced by the feedback factor. At intermediate points, they are both amplified and canceled. The gain to an

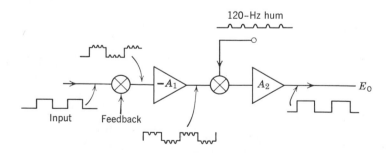

Figure 4.24   Hum cancellation in a feedback amplifier.

internally derived signal was discussed in Section 4.13 and is

$$\frac{E_O}{E_C} = -\frac{1}{A_1\beta}$$

where A_1 is the gain preceding the point of injection.

4.22   FEEDBACK AND SIGNAL LEVELS

As shown above signal levels within a feedback loop are directly related to the gain distribution. If the output signal is known, any other internal signal level can be calculated. When a frequency-sensitive gain element is used, the signal level ahead of the element adjusts to provide the correct output as a function of frequency. Such a change in signal level can be a source of trouble. If the input stage swings 0.1 V at a low frequency and there is a 100:1 high-frequency gain reduction following this point, the signal level at the input stage must increase by essentially 100-fold to provide the required output. This increase to 10 V may be impractical because of power-supply limitations, non-linearity in gain, or the inability of a stage to provide current for reactive elements.

To avoid unnecessary difficulty, frequency-dependent gain changes should be located at or near the input stage of an amplifier. With this approach the only signal to increase is the error signal. Since input-stage signal levels are small, the reactive currents required are minimum. It is important that no more signal loss be taken than is required unless bandwidth and output signal level are unimportant. Note that if an input stage is caused to operate in a nonlinear region, the resultant errors are reduced by the feedback factor and multiplied by the gain factor following.

4.23   FEEDBACK FACTOR AND NOISE

Noise sources can be reduced to an equivalent noise-voltage generator injected into the signal path. Noise generated by the input stage is essentially an input signal and for this discussion the gain to the input noise is simply $1/\beta$.

When a second gain stage generates noise, this noise is amplified by $1/\beta$ and is reduced by the gain preceding the point of injection A_1; that is,

$$E_O - \frac{E_N}{\beta A_1} \tag{20}$$

If the noise generated in a second stage is 100 μV and the input stage gain is 100, the output has a noise level of 1 μV that appears to come from the input.

When the input stage is attenuated at high frequencies, the denominator in (20) reduces and the noise contribution of a second stage at high frequencies increases. If the first stage has a gain of 1, it is obvious that both the first and second stages can contribute equally to output noise.

Second-stage noise is often higher than input-stage noise and the input-stage gain is important if second-stage noise is to be masked. The penalty for shaping the frequency response in an input stage is thus the inclusion of second-stage noise. If this solution to a stability problem is used, then the second stage should also be low noise. If the frequency shaping occurs in the second stage, the increased signal levels required of the input stage must then be considered.

PROBLEMS

1. In the operational feedback structure of Figure 4.1 $R_1 = 10$ kΩ and $R_2 = 100$ kΩ. If the input current is 10 μA from a 1-MΩ source, what is the output voltage?
2. The amplifier in Figure 4.1 injects 10 nA into the point S. What is the output voltage if the input terminals are shorted? If they are open?
3. In Problem 1, if the amplifier's open-circuit input impedance is 100 kΩ and a 1-V input signal is provided, what input current flows in the input impedance?
4. Using Figure 4.1, if $R_1 = 10$ kΩ and $R_2 = 100$ kΩ, consider a 1-V input signal and calculate the voltage at S. Using the input current in R_1, calculate the impedance at S.
5. Using Figure 4.4, if $R_1 = 10$ kΩ and $R_2 = 5$ kΩ, calculate the voltage gain. If R_1 is changed 1%, how much does the gain change? Repeat for R_2.
6. What is the value of β in Problem 5?
7. An amplifier has an input R of 100 kΩ and delivers power into a 100-Ω load. If the voltage gain is 1000, calculate the true power gain in decibels. What is the voltage gain in decibels assuming a common impedance level?
8. In Figure 4.1 $R_1 = 10$ kΩ and $R_2 = 100$ kΩ. Assuming the amplifier has zero gain and the output Z is 100 Ω what is the direct-transmission gain? If R_2 is shunted by 100 pF, at what frequency is the direct transmission increased by 6 dB?
9. In Figure 4.11 show that the distance to the A curve from the line $1/\beta = 40$ dB is simply $A\beta$.
10. In Figure 4.12 if $R_1 = 10$ kΩ, select values for C_2 and R_2 so that the transmission is reduced 3 dB at 10 kHz and has a final attenuation factor of 40 dB. At what frequency is the attenuation factor 37 dB?
11. If the parasitic collector capacitance in Figure 4.12 is 100 pF, compute the two 45° phase shift frequencies in Problem 10. Repeat for $R_2 = \infty$. Plot the phase shift versus frequency for Problem 10.
12. In Figure 4.16 the step response is given by $T_0 = E_{IN}e^{-t/RC}$. Let $C = 1$ μF and $R = 1000$ Ω. Assume a sinusoidal input; what is the period of one cycle in

which the attenuation is 3 dB? Compare this with the time to sag to the 70% point for the step function.

13. An output coupling capacitor of 10 μF is used with a dc amplifier. What load resistor is permitted which attenuates the signal 1% at 20 Hz?

14. In Figure 4.23 assume the β block is nonlinear. Sketch the waveforms at all points in the circuit.

15. An input stage has a 10-μV noise level and a gain of 5. A second stage has a noise level of 50 μV. What is the total noise referred to the input? (Remember noise power adds.)

5

Feedback Amplifiers (continued)

5.1 GAINS AND OPERATIONAL FEEDBACK

The gain of a feedback amplifier is determined by its feedback structure or β loop. In many practical instruments, it is desirable to switch gains by modifying the feedback elements. If a continuous gain adjustment is desirable, potentiometers can be used in the β loop. The next sections discuss a few standard techniques useful to the designer.

Operational feedback uses the summing principle at the input. If there is adequate feedback, the summing point is a virtual ground point; that is, the potential at the summing point is essentially zero. The signal currents flowing into and out of the summing point are equal, and because of the lack of signal potential almost no signal current flows in the input impedance of the amplifier.

This required null at the summing point provides a useful device for quickly calculating the ideal gain of a feedback structure. An example is given in Figure 5.1. The signal at S is zero; therefore the signal at E_1 is known. (The current flowing into R_1 is equal to the current flowing in $10R_1$, and $E_1 = 10E_{IN}$.) This ratio of E_1/E_{IN} is independent of resistor values R_2 and R_3. The overall gain from input to output is determined by asking the question: What output signal is required to produce the voltage E_1 at P? At first glance, if $R_2 = R_3$, E_0 is twice E_1, and the overall gain would be 20. It is necessary to consider that $10R_1$ is in parallel with R_3 as the voltage at P must supply current

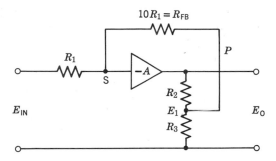

Figure 5.1 A feedback arrangement using an attenuator.

to both resistors. Thus, if R_2 equals the parallel resistance of R_3 and $10R_1$, the gain is 20. Stated as an equation, where primes indicate loaded values,

$$E_O = -E_{IN} \frac{10R_1}{R_2} \frac{R_2 + R_3'}{R_3'}$$

where

$$R_3' = \frac{R_3 \times 10R_1}{10R_1 + R_3}$$

The lowest gain possible by varying R_2 or R_3 is 10. If gains below 1 are required, then the feedback element R_{FB} must be reduced to a value below R_1. Operational feedback amplifiers *do* permit gains below unity without the use of an input attenuator. This feature is not available in potentiometric feedback.

The value of R_3 should never be zero; if it is, the feedback loop is open and the amplifier gain goes to the open-loop value. A typical lower-limit value for R_3 is 10 Ω. If precision gains are required, this resistor is usually four-terminaled and/or specially switched to reduce end resistance effects. If lower values seem necessary, a ladder output attenuator is suggested. This technique is preferred over using a higher impedance attenuator as such an element is more subject to parasitic capacitance effects which in turn reduces high-frequency performance. If wide-band response is unnecessary, then the higher values can be used.

5.2   VARIABLE GAIN

When resistor R_{FB} is varied on a continuous basis, is can constitute a variable-gain element. If the attenuator made up of R_2 and R_3 is unaffected by the loading of R_{FB}, the variable gain is linear. If $R_2 = 0$, loading is not a problem and the control is exactly linear. This control can accommodate a counting dial to indicate gain numerically. If the resistor R_{FB} is a potentiometer, the circuit has the following problems:

1. One side of the potentiometer can be at the summing point, a critical point to extend in dimension.

2. The full output signal is brought out to the potentiometer, adding to the capacitive coupling to surrounding circuitry.

3. The impedance level is usually high to keep the input impedance high. High-value precision wirewound resistors are undesirable because of parasitics and reliability problems. Parasitics can cause the frequency response to be a function of control setting.

4. The resistance can cover only a range that corresponds to stable amplifier operation; that is, the range must be limited to that in which the Nyquist stability margin is adequate.

If the range of R_{FB} is limited, the limit resistor is best placed on the summing-point side. A span of 10:1 is a recommended maximum. Typical resistance values might be a 6.66 kΩ fixed R and a potentiometer of 10 kΩ to cover the gain multiplication range from 1 to 2.5.

Resistor R_3 can also be used as a variable-gain control. Since it is desirable to use one control-element value, the attenuator impedance level can be changed for each gain range. Assuming a gain 100 and no attenuator loading,

$$\frac{R_2 + R_3}{R_3} = 10 \tag{1}$$

At gain 200, assuming R_3 changes to R_4,

$$\frac{R_2 + R_4}{R_4} = 20 \tag{2}$$

If $R_3 - R_4 = 1$ kΩ, eq(1) and (2) can be subtracted to show that $R_2 = 17.1$ kΩ, $R_3 = 1.9$ kΩ, and $R_4 = 0.9$ kΩ.

This gain arrangement in Figure 5.2 is satisfactory as long as the resistance change does not load the amplifier unnecessarily. To design a variable-gain circuit that leaves the load constant, the total attenuator resistance must be held constant. If R_P is the potentiometer resistance and if R_{FB} does not load the output attenuator, then $R_2 + R_3 + R_P = R_T$. At gain 100 the attenuator has a loss of 10 and at gain 200 it has a loss of 20. The ratios of resistors are

$$\frac{R_T}{R_3} = 20 \quad \text{and} \quad \frac{R_T}{R_3 + R_P} = 10 \tag{3}$$

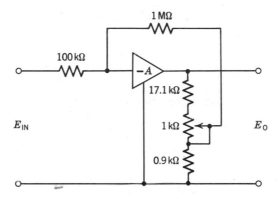

Figure 5.2 A 2 : 1 variable-gain control.

If $R_P = 1\ \text{k}\Omega$, then $R_3 = 1\ \text{k}\Omega$, $R_1 = 20\ \text{k}\Omega$, and $R_2 = 18\ \text{k}\Omega$. The circuit is shown in Figure 5.3. The loaded attenuator case is not as simple to calculate. If the loaded values R_3' and $(R_P + R_3)'$ are held to 1 kΩ and 2 kΩ, respectively, the gains will be correct. (In the example above the 1-MΩ load affects the gain by only 0.1%) If this loading error must be considered, it is practical to start by assigning a value to R_P. If loaded values are noted by primes, the following equations can be used, in which G_1 and G_2 are the two gain values:

$$\frac{R_T}{R_3'} = G_1 \qquad \text{and} \qquad \frac{R_T}{(R_P + R_3)'} = G_2$$

where

$$R_3' = \frac{R_3 R_{\text{FB}}}{R_3 + R_{\text{FB}}} \qquad \text{and} \qquad (R_3 + R_P)' = \frac{(R_3 + R_P)R_{\text{FB}}}{R_T}$$

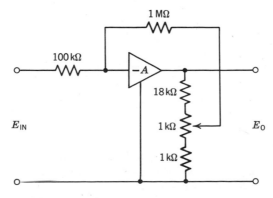

Figure 5.3 A variable gain with constant amplifier load.

These equations can be manipulated to show that

$$\frac{(R_3 + R_{\text{FB}})^2(R_3 + R_P)}{R_3{}^2 R_{\text{FB}}} = \frac{G_1{}^2}{G_2} \tag{4}$$

where R_{FB} and R_P are given values. Equation (4) is a cubic which is best handled by approximation techniques. Once R_3 is known, R_4 and R_2 can be found by a simple calculation.

The input resistor R_1 can be changed to affect gain, but a change here modifies the input impedance. Also note that the resistance is reciprocally related to gain. These factors are usually undesirable and this approach is used only for variations of a few percent. Such small variations are often required for gain trims. Here, the control is placed at the input terminal rather than at the summing point. If gain trims are placed in other parts of the feedback structure, they additionally complicate the accuracy problem. When R_1 is varied, the gain change is independent of other gain changing resistors.

5.3　NOISE CONSIDERATIONS

The input resistor R_1 and the feedback resistor R_{FB} both contribute to the output noise. The noise generator associated with R_1 is in series with R_1, and the gain to the output is just the expected gain of the feedback amplifier. A 100-kΩ resistor (R_1) is a noise generator of 5.6 μV rms for 20 kHz bandwidth. If the gain is 1000, this value appears as 5.6 mV rms in the output.[1]

If R_2 is greater than R_1, the noise contribution from R_2 can usually be ignored. If $R_2 = 3R_1$, the noise increase is usually not more than 1 dB. It is of interest to calculate the noise contribution from R_2 in a typical case. In Figure 5.4 the current supplied by E_{R2} to the summing point must be zero. This requires that $E_1 = -E_{R2}$. The value of E_0 for this source is therefore $-E_{R2}(R_3 + R_4)/R_4$. The noise voltage at the output resulting from E_{R2} is independent of R_1, whereas the noise voltage at the output from E_{R1} is directly related to R_2.

In the example above, if $R_2 = 4R_1$ and if $(R_3 + R_4')/R_4' = 2$, the noise generator E_{R1} appears as $8E_{R1}$ at the output because the amplifier gain is 8. Noise generator E_{R2} appears in the output as $2E_{R2}$. Since noise voltages are not additive, it is necessary to consider noise power that can

---

[1] A factor of approximately 6, converts rms volts to peak-to-peak volts since the form factor for random noise is not the same as that for sinusoids. In the example the output noise would be 35 mV peak-to-peak.

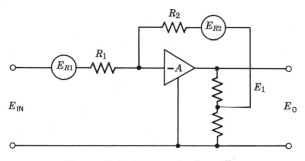

Figure 5.4 Noise gain from E_{R2}.

be added. Noise power is proportional to the noise voltage squared; thus the total noise power P_N is

$$P_N = K(E_1{}^2 + E_2{}^2) = KE_T{}^2 \qquad (5)$$

where E_1 and E_2 are output noise voltages resulting from E_{R1} and E_{R2}, respectively, and K is the proportionality factor. Thus E_T, the combined noise, is

$$E_T = \sqrt{E_1{}^2 + E_2{}^2} \qquad (6)$$

In the example above, using the calculated values of E_1 and E_2, we have

$$E_T = \sqrt{(8E_{R1})^2 + (2E_{R2})^2}$$

From (1) in Section 1.1 the noise voltage is proportional to the square root of resistance. In this example $R_2 = 3R_1$ and thus $E_{R2} = \sqrt{3}\, E_{R1}$. When using this value for E_{R2}, the total noise is

$$E_T = \sqrt{64E_{R1}{}^2 + 12E_{R1}{}^2} = E_{R1} \times 8.70$$

The noise generator E_{R2} thus adds to the noise output in the ratio 8.70 to 8 or 9%. This is approximately 0.9 dB.

In low-level dc amplifier designs the noise contribution from R_1 is undesirable. Since R_1 defines the input impedance and the noise, high values of R_1 will not fit the need. This conflict in requirement is usually resolved by using potentiometric feedback. In some operational feedback applications, the current flowing into R_1 is supplied from a current source. This current might result from a phototube, a photomultiplier, or a semiconductor. In these applications R_1 is internal to the device supplying the current. The noise contribution here is a part of the source mechanism. A typical phototube input is shown in Figure 5.5.

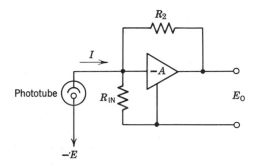

Figure 5.5 A phototube input to an operational amplifier.

Because of the amplifier feedback, little voltage appears across R_{IN}; thus most of the signal current flows in R_2. The gain of this circuit is the ratio of output voltage to input current; that is, $G = E_0/I_{IN}$. Since $E_0 = I_{IN}R_2$, the gain $G = -R_2$. It is correct to say this amplifier has a gain (transimpedance) of 1 MΩ. By Ohm's law 1 μA of input current will produce 1 V of output signal.

One low-noise technique employing an operational feedback circuit uses the source resistance as the input resistor. This approach limits the noise contribution to the source resistance itself but removes the open-circuit voltage from any secondary observation. This circuit is shown in Figure 5.6.

In effect the operational amplifier measures the short-circuit current available from the source. (The point S is at null.) In many cases the short-circuit current from a transducer is just as linear as the open-circuit voltage. Because of the use of signal power, the former approach can often yield a better S/N ratio.

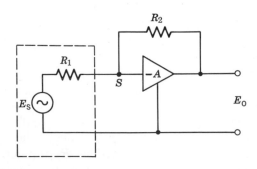

Figure 5.6 Operational feedback, using the source R_1 as a feedback element.

Noise currents entering the summing point S from within the amplifier cannot be distinguished from signal currents entering through the input resistor. The gain to these currents is therefore the same as the gain to signal currents. Most noise specifications are related to an input signal voltage. For an operational amplifier this voltage is dependent on the value of R_1. If R_1 is large enough, the resistor noise will be much greater than the amplifier noise.[1] If the noise of the amplifier itself is to be considered, the only useful measure is in terms of input noise current. To measure this noise current, the amplifier should be operated using low-value feedback resistors to eliminate their contribution. The resulting output noise E_N can be related to an input noise current I_N by Ohm's law. If R_1 is the input feedback resistor, then

$$I_N = \frac{E_N}{GR_1} \tag{7}$$

where G is the amplifier gain.

All currents entering the summing points other than signal currents are undesirable. The noise currents considered above cover the spectrum from zero frequency to the band of interest. The static component or dc value is usually singled out in any discussion although technically it is just one part of a total noise picture. The reason for this differentiation involves offset and drift considerations. Most of the dc component of noise current can be subtracted by suitable circuit technique, whereas the fluctuating aspects cannot.[2]

5.4   GAINS IN POTENTIOMETRIC FEEDBACK

Potentiometric feedback always involves two input points. These points have nearly equal open-loop signal gain to the output terminal but they are of opposite sign. The standard input configuration involves a balanced pair of transistors usually operated in a symmetrical manner from one current source. (The reader should review the material in Section 4.7, in which these ideas are first discussed.) The output attenuator shown in Figure 5.7 provides a convenient method for switching the

---

[1] This assumes that other noise phenomena such as power-supply ripple, lack of regulation, spikes, and rf hash are not involved.

[2] Philosophically, fluctuations occurring once per day are still alternating current. This simply means that static compensation accommodates only fixed operating conditions. If variations are present, they are classed as alternating current. If the static value shifts in value, it is still alternating current although accepted practice is to say that "the dc level changed." This conflict is a semantic one only. Once the mechanisms are understood, the physical facts are the only problem.

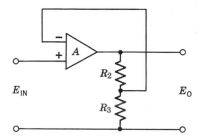

Figure 5.7   A potentiometric gain configuration.

gain. The voltage gain is positive and is given by the ratio

$$G = \frac{R_2 + R_3}{R_3} \tag{8}$$

Viewed another way, the output voltage is just that signal value which when attenuated feeds back a signal equal to E_{IN}.

5.5   GENERAL DISCUSSION

The input terminals to a potentiometric amplifier both move with the signal as opposed to the operational case, in which the summing point is a virtual ground. If the signal gain is high, the difference voltage at the input must be small. If the amplifier gain is unity, the input signal points have the same signal levels as the output. This requirement places two severe demands on the circuitry. Parasitic capacitances require signal current, and the characteristics of the input transistors must remain matched over the full signal operating range.

The signal voltage levels at the input terminals of a potentiometric device are limited by the operating range of the input stages. This limitation is not present in operational feedback when the summing point is at a null. Operational feedback places no restriction on voltage levels as long as a current balance occurs. Another limitation in potentiometric feedback involves gains below unity. If gains below one are required, an input attenuator is necessary. Attenuators that maintain the high input impedance are subject to two difficulties: Source currents flowing in high-valued resistors can upset the amplifier's operation, and component parasitics can reduce the dynamic performance below acceptable limits.

5.6   INPUT IMPEDANCE

The chief advantages of potentiometric feedback are the low noise and the high input impedance. The low noise results from the fact that no series operational resistor is necessary. The high input impedance results from feedback returned to the input differential pair. The feedback reduces the flow of signal current into the input resistance, and the reduced current flow is a direct measure of input impedance.

The transfer characteristics of the differential-input-transistor stage are directly involved in the input impedance calculation. In Figure 5.8 the current source I_C is assumed to be perfect; that is $I_1 + I_2 = $ constant. The output voltage is just the difference between collector potentials, or $I_1 R_1 - I_2 R_1$. Without an input signal $I_1 = I_2$ and the output voltage difference is zero. The emitter-base potential difference of the input transistor is $E_1 - E_B$. After a small change in E_1 the base moves to a new potential, $E_B + \Delta E_B$. The change in base-emitter potential difference is $[(E_1 + \Delta E_1) - (E_B + \Delta E_B)] - (E_1 - E_B) = \Delta E_1 - \Delta E_B$. In Figure 5.8 the second transistor has a fixed base potential and the voltage ΔE_B subtracts from the initial potential difference after ΔE_1 is applied. The

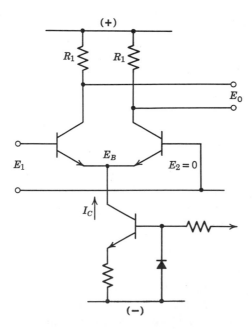

Figure 5.8 A differential input stage.

transconductance g for each transistor is defined as the ratio of base-emitter potential difference to collector current difference, or

$$g = \frac{\Delta I}{\Delta E}$$

The current in the input transistor changes to

$$I_1 + \Delta I_1 = I_1 + g\,\Delta E = I_1 + g(\Delta E_1 - \Delta E_B) \tag{9}$$

For the second transistor the new current is

$$I_2 + \Delta I_2 = I_2 + g\,\Delta E = I_2 - g\,\Delta E_B \tag{10}$$

The output difference of potential is thus

$$E_O = [I_1 + g(\Delta E_1 - \Delta E_B)]R_1 - (I_2 - g\,\Delta E_B)R_1 \tag{11}$$

Since $I_1 + I_2$ is constant, the current change in the two transistors must balance; thus $g\,\Delta E_B = (\Delta E_1 - \Delta E_B)g$ or $\Delta E_1 = 2\Delta E_B$.
Using this result and noting that $I_1R_1 - I_2R_1 = 0$ yields, from Equation (11),

$$E_O = g\,\Delta E_1 R_1 \tag{12}$$

From (9), the change in input collector current is $g\Delta E_1/2$. If the ratio between collector current and base current is β, the change in input base current is $g\Delta E_1/2\beta$. The input impedance Z_i is just $\Delta E_1/\Delta I_1$, and Z_i becomes

$$Z_i = \frac{2\beta}{g} \tag{13}$$

In a typical input stage $\beta = 50$, $g = 0.01$ mho, and $Z_i = 5000\ \Omega$.

This impedance level can be increased by feedback to perhaps 5 MΩ. Such a value is not adequate for some differential amplifier designs in which 1000 MΩ is required. To achieve these higher values the effective β of the input stage must be at least 10,000. Such a value can be attained by using an input Darlington connection which multiplies the β of two transistors. Such a circuit is shown in Figure 5.9.

The preceding analysis was based on a grounded base connection for the second transistor. Equation 13 was derived assuming that the change in input base-emitter potential resulted entirely from a signal applied to the input base. To a very close approximation, the dynamic input impedance of the circuit in Figure 5.8 is independent of the static base potential on the second transistor. This assumes that the transistors are operating within their normal limits.

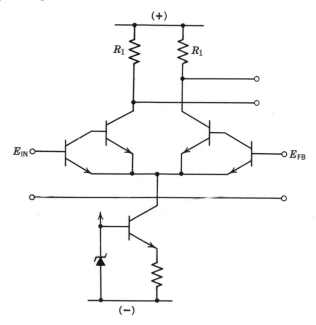

Figure 5.9 A Darlington connected differential stage.

If both input bases are changed in potential by the same voltage, the common-base voltage will closely match this change. The change in base current required to effect this change is negligible as both collector currents remain constant. The input impedance under this condition is essentially infinite. In a practical feedback scheme the feedback signal returned to the second base only approaches the value of the input signal, and the input impedance reflects this fact. The dynamic input impedance is related to changes in potential difference between the two input bases rather than to the potential changes between the input base and input emitter.

In practice many second-order effects enter into a full calculation of dynamic input impedance. From the preceding discussion it should be apparent that very careful parameter matching is an essential part of keeping the input impedance high. When a large range of operating temperature is required and when a range of supply voltages is expected, parameter matching becomes even more complex. The subject area is treated rather extensively in the literature as it is the essence of quality performance in low-level *IC* amplifiers and low-level instrumentation amplifiers.

The β loop of a potentiometric feedback scheme includes the input stages of the amplifier. If these active elements misbehave, the performance can *not* be changed by adding more loop gain. More feedback only tends to ensure that the incorrect quality entrenched in the β segment be more closely followed. If the second gain element of the input transistor pair has a reduced gain at high frequencies, the output will peak in response at high frequencies. This corresponds exactly to the statement that the gain approaches $1/\beta$ for $A\beta >> 1$.

In a normal feedback loop using potentiometric feedback it is very difficult to measure stage gain by observing signals at the collectors of the input stages. Even though the input bases can have large signals, the collector differences of potential are usually very small. Signal gains can usually be inferred by operating the amplifier in open-loop segments at reduced signal levels. If the closed-loop performance is upset by a stability problem, all rules are off for making useful gain measurements. Care should be taken to ensure that the presence of measuring equipment does not add overloading pickup or create instability. Confidence in internal closed-loop measurements comes only with experience. An output monitoring oscilloscope is a necessity in making these observations. It is important to realize that internal measurements made near the input stages of a high-gain feedback amplifier are most apt to upset the amplifier operation and result in misleading information. Note that *dc* measurements can be in error if the observation upsets the amplifier's operation.[1]

5.7 NOISE IN POTENTIOMETRIC AMPLIFIERS

The dominant noise source to consider in a potentiometric feedback scheme is a voltage generator in series with the input signal line. Such an equivalent generator can be measured by observing the output noise for a short-circuited input and dividing the resultant output signal by the amplifier gain.

The noise from each input transistor contributes equally to the total noise in a power sense. The balanced structure is thus not ideal from a noise standpoint, but this disadvantage is far outweighed by the other qualities of the configuration.

The noise contribution of feedback attenuator resistors is small enough to be neglected. The noise from high-valued source resistors, however, is not in the same category and must be considered. The internal noise

[1] A high-value resistor added to the meter probe at the point of observation usually limits the parasitic loading of the meter.

current flowing in the source resistance produces additional noise. Input transistors can be operated at current levels which optimize the total noise as observed at the output. Component specifications cover this area thoroughly particularly when the elements are intended for "front-end" application.

5.8   INPUT IMPEDANCE WITH POTENTIOMETRIC FEEDBACK

The input impedance of a differential stage without feedback is discussed in Section 4.8. When the return base is connected to a fraction of the output signal, the gain of the system becomes the reciprocal of this fraction. The presence of this feedback signal reduces the signal current required from the signal source. The reduction in current raises the input impedance.

In a closed-loop system the collector potential difference at the input differential transistor pair is defined by the gain following. If the gain following is doubled, the collector difference signal must halve. This lower signal level reduces the signal current demand from the input source. If the open-loop collector difference of potential is 100 mV before feedback and 0.1 mV after feedback, the input impedance is multiplied 1000-fold by the feedback. This multiplication argument is not completely valid for very high input impedances. As indicated before, second-order differences in the differential pair limit the maximum impedance level possible because the source must supply current to compensate for these differences.

Input impedance and input current are two separate specifications. Input impedance is a dynamic specification only. It is the ratio of signal voltage *change* to signal current *change*. Large static currents may be undesirable, but they are not a part of an input impedance specification. Current changes are often less than static current levels. This subject is discussed further in Sections 7.15 and 8.19.

Potentiometric feedback does not raise the input impedance resulting from elements paralleled across the input terminals. Input cable capacitances or a resistor to signal common simply parallels the dynamic input impedance discussed above. The input capacitance due to the transistors is modified, however. The constant-current source ensures that the collectors move only differentially. With sufficient feedback, the collectors remain essentially fixed, and the only charging current that flows involves the collector-to-base capacitance. The capacitance multiplication effect that results from collector potential change (Miller effect) is then not present. The base-collector capacitance shunts the input terminal, and this is a limiting effect in the high-frequency input impedance of a potentiometric feedback amplifier.

5.9 VARIABLE GAIN AND POTENTIOMETRIC FEEDBACK

The variable-gain discussion under operational feedback provides the basis for variable gain in potentiometric feedback amplifiers. The output attenuator is not loaded by the feedback connection, and thus the gain for the circuit in Figure 5.10 is given by the expression

$$G = \frac{R_2 + R_P + R_1}{R_2 + (1 - x)R_P} \tag{14}$$

where x represents the fraction of R_P on one side of the slider.

In some amplifier designs it is required to consider the variable gain in reciprocal units; for example, the gain might be in units of "mV to produce full scale output" or mV/V. A lower number of mV on the indicator represents a higher voltage gain. From (14) the reciprocal of gain, called S, is linear in x; that is,

$$S = \frac{\text{mV}}{\text{V}} = \frac{R_2 + R_P}{R_2 + R_1 + R_P} - \frac{R_P x}{R_2 + R_1 + R_P} = A - Bx \tag{15}$$

Since A and B are constants, values of resistance can be found that permit S to linearly span two values.

If a reduction in input impedance is acceptable, an input potentiometer can serve as a variable gain in a potentiometric amplifier. If the amplifier has a high input impedance, the potentiometer is unloaded and provides linear gain *to zero* if desired. The input impedance is a constant and

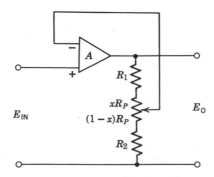

Figure 5.10 Variable gain and potentiometric feedback.

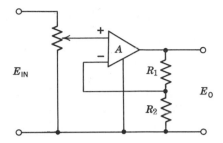

Figure 5.11   An input attenuator.

is simply the value of the input potentiometer. This configuration is shown in Figure 5.11.

Referred-to-input noise is raised by the input attenuation factor. If large signals are involved, this consideration is unimportant. As a general rule, input attenuators are not used in instrumentation. It is preferable to amplify a signal before taking a signal loss to preserve the S/N ratio.

5.10   OFFSET CIRCUITRY AND ZERO CORRECTIONS

Dc amplifiers all require zero adjustment. Circuits that modify the zero may accommodate internal zero errors or external signal requirements. The internal errors can result from either current sources or voltage sources. The external signal requirements can exist at the input or at the output.

Many adjustments are made once and are not changed after manufacture. These zero corrections involve second-order effects in which a small current or voltage is inserted into the circuitry. Zero corrections are often made temperature-dependent to compensate for internal temperature-dependent errors. Some adjustments cover a limited range and use a variable-control element that can be changed with a knob or screwdriver. When the zeroing element covers a large range, it is often referred to as an offset control.

Zeros or offsets can be referred to either input or output points. When offsets are referred to the input, the abbreviation used is RTI. Similarly the output offset is abbreviated RTO. It is possible to have zero effects that require compensation at both the input and output points. Two corrections are usually required if the gain of the circuit is varied by feedback changes. Techniques vary depending on the type of feedback, the type of correction, and whether the correction is RTI or RTO.

5.11   TESTING RTI VERSUS RTO ZERO CORRECTION

The test of RTI or RTO condition can be made by changing the circuit gain. If a gain change leaves the output offset fixed, the offset is RTO, whereas, if a gain change tracks exactly with the zero error, the zero error is RTI. Obviously zero offsets of both types can exist simultaneously. One method of separating variables is to observe the offset while changing or "rocking" the gain. When a zero adjustment can be found that leaves the output zero constant for gain changes, the remaining zero error is RTO. This does not imply that all zero errors that are RTI are then subtracted. Input zero errors can result from both input current and input stage unbalance. A single point adjustment thus can correct for several separable zero errors. A change in an input parameter can change the zero RTI unless the errors are separated and individually balanced out (see Section 5.15).

5.12   RTI ZERO ADJUSTMENTS—OPERATIONAL FEEDBACK

Currents added to the input summing point can be used to compensate for internally supplied current sources. A large resistor connected to a regulated voltage can provide the required current. A typical circuit is shown in Figure 5.12. If $R_3 >> R_1$, the noise RTI is not affected by R_3. If E_{REG} changes, the output zero will vary with this voltage. If R_3 is varied, the output is changed over a range of negative voltages. As an example, if $R_2/R_1 = 100$ and $R_2 = 1$ MΩ, then a current of 10 μA flowing into the summing point produces -10 V of output offset. If $R_3 = 100$ MΩ, the offset from a 10-V regulated source is -0.1 V. Since zero adjustments often compensate for a unipolar current source,

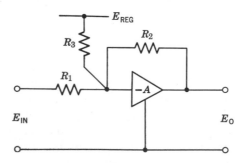

Figure 5.12 Current correction in an operational amplifier.

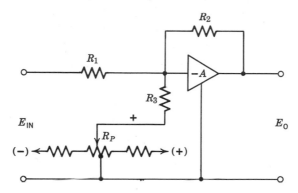

Figure 5.13   A practical operational feedback zero circuit.

one polarity of current source is usually adequate.[1] Varying R_3 is not a recommended procedure for adjusting zero; a preferred arrangement is shown in Figure 5.13. The voltage on the potentiometer slider can be varied over a bipolar range of voltage. If $R_3 >> R_P$, no loading occurs and the offset is proportional to potentiometer setting.

The potentiometer in Figure 5.13 is shown center tapped. When the slider is at the center, variations in individual supply voltages cannot introduce a current into R_3. If the supplies are adequately regulated, the center tap may not be required.

If the potentiometer spans only a limited voltage range, a pseudo center tap can be formed that attenuates any power-source variation. Such a circuit is shown in Figure 5.14. If this circuit is used to produce full-scale output offsets, these offsets can be switched and ganged to gain changes so that they appear constant at the output or, in effect, appear RTO. If $R_3 = 100$ MΩ[2] at gain 100 and 50 MΩ at gain 50, the RTO signal is constant with gain—under the assumption that there is no variable gain.

If R_1 is varied to produce a variable gain, the output zero is not affected as offset is a ratio between R_3 and R_2. If R_2 is varied, any offset introduced by R_3 will vary proportional to R_3. Thus, if R_3 is "gang-switched" with gain to effect an RTO offset control, R_1, not R_2, must be used for a variable-gain control.

---

[1] The source voltage can be a diode. First-order temperature compensation results as the diode voltage drop and the source current vary together. The lower voltage will result in lower values of R_3.

[2] Resistor values above 22 MΩ are not standard stock. If the reference voltage is reduced to 0.1 V, the resistor values required will be reduced in value by two orders of magnitude.

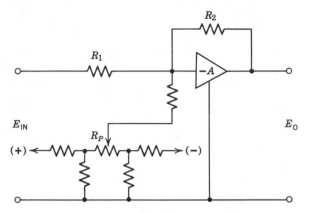

Figure 5.14 A pseudo center tap for zero control.

The zero stability of an offset circuit is as critical as the zero drift of the amplifier. The circuit in Figure 5.13 adds no zero errors at a zero setting. Ideally the zero drift errors introduced at full-scale offset should not exceed the zero drift errors of the amplifier without offset. In practice this requirement is difficult to meet.

5.13 OPERATIONAL FEEDBACK OFFSET REFERRED TO OUTPUT

RTO offset circuit can be added to the β structure of an operational amplifier. This configuration is shown in Figure 5.15. The point S can be considered a second summing point for currents flowing from node E_S. The point S_1 is always near zero signal volts, and the current in R_1

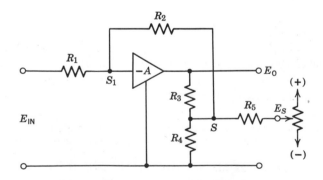

Figure 5.15 An RTO offset circuit.

equals the current in R_2. If the current in R_1 is zero, the current in R_2 is zero, independent of all other considerations. If a current flows from E_S, it cannot flow in R_2 as the point S remains at 0 V. The only path for current entering S is through R_3. Thus $E_0/R_3 = E_S/R_5$ or $E_0 = E_S \times R_3/R_5$, the gain relationship of an operational feedback amplifier.

If R_4 is varied, either in steps or in a variable-gain manner, the offset voltage at the output is constant. This feedback arrangement is ideal for galvanometer positioning when a mechanical zero can be electrically accommodated. Note that R_3 must be finite and that the loading of R_5 and R_4 must be considered for a gain determination. If $R_3 = 0$, no offset results.

5.14 POTENTIOMETRIC FEEDBACK OFFSET CONSIDERATIONS

The potentiometric circuits described before have an input base and a second or feedback base. The feedback process ensures that the difference of potential between these bases is near zero. If the input stages are slightly unbalanced, the output signal will mirror the unbalance when the feedback loop is closed. This characteristic is demonstrated in Figure 5.16 in which a 1.5-V battery is placed in series with the feedback base. By previous discussions the two input bases must ideally be at the same potential. This requires that the feedback signal be equal and opposite in voltage to the battery voltage E_B. The output voltage must also overcome the output attenuator loss ratio. The output voltage is equal to $E_B(R_1 + R_2)/R_2$, that is, E_B times the signal gain. Thus voltages added in series with the feedback base cause an offset signal RTI.

A floating voltage source as in Figure 5.16 is impractical and is rarely

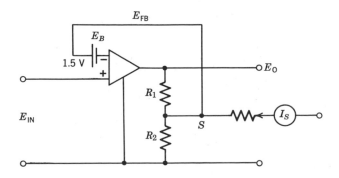

Figure 5.16 Offset in potentiometric feedback circuits.

used for RTI offset. It does illustrate the point that the output voltage in a closed-loop system is directly related to the unbalance existing in the input differential stages.

Offset current can be summed into the point S as in the operational circuit of Figure 5.16. The feedback signal E_{FB} does not change which implies that the current into S flows in R_1 or $E_0 = -I_S R_1$. If R_2 is changed, the offset remains constant RTO. If R_1 is changed, the offset and gain are both affected but they do not track. Note that $E_0 = -I_S R_1$, whereas $E_0 = E_{\text{IN}}(R_1 + R_2)/R_2$ for an input signal E_{IN}.

5.15   INPUT CURRENT EFFECTS OR "PUMPOUT"

In all semiconductor devices, some gate or base current flows.[1] If FETs are involved, input currents will be in the picoampere range. For transistors this current is usually measured in nanoamperes. When an output zero error is exhibited by the operational feedback circuitry of Figure 5.14, it can be canceled by summing external currents into the summing node. This method can correct for input voltage errors as well as for currents being "pumped" back from the input transistor.

Zero errors that are attributable to source-current flow can be sensed by varying the input source impedance. If the output varies as a function of source impedance (say, from 0 to 1 kΩ), the variation indicates the presence and magnitude of the phenomenon. If the gain is 1000 and a 1 kΩ change produces a 10-mV output shift, the current is 10 μV in 1 kΩ, or 10 nA. When this current is supplied from an external source, the output voltage will not vary as a function of source resistance.

The output voltage is not necessarily zero when the flow of "pump-out" current is removed from the source impedance. To zero the output a second zeroing is required. The zeroing technique can take on the form shown in Figure 5.17. Resistor R_P supplies input base current so that no static base current flows in R_S. Variable voltage E_V is supplied to the base of a differential input pair to ensure that zero input voltage at S produces zero output voltage at E_0. Procedures for setting R_P and E_V are essentially independent of each other. If E_P is a large voltage, small variations in potential at S will not alter the current in R_P. A value determination of R_P is usually made during manufacture and requires changing only if new input transistors are required.

Transistor base current flowing into S varies as a function of temperature. A circuit such as that in Figure 5.17 for supplying this current is thus valid at only one temperature. A first-order compensation for tem-

---

[1] This current is sometimes called "bias current."

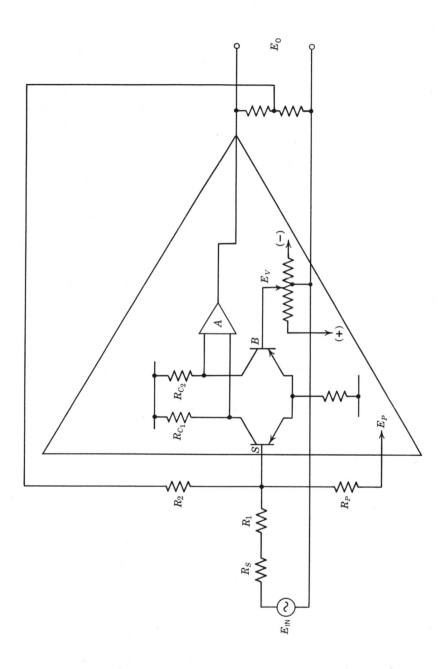

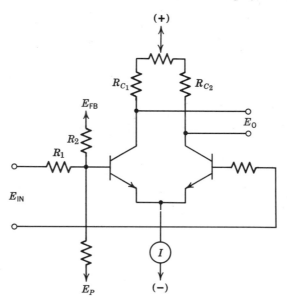

Figure 5.18   A collector zeroing circuit.

perature-sensitive offset involves adding a series R between transistor base B and the voltage source E_V. If this R is equal to the resistance as seen by base S, then both base potentials change equally with base temperature. The drift is thus balanced out as $E_S - E_B$ remains constant. This form of compensation is adequate for low-impedance sources R_S. The current in R_P can be made a function of temperature by varying E_P as a function of temperature, that is, a diode source. Source impedance zero dependency can be removed only if the source current in R_S is zero as a function of temperature.

The zeroing technique by which E_V is varied is ideal. Other internal voltage adjustment schemes are also possible. Resistor R_{C1} or R_{C2} can be changed to introduce a difference signal. A balanced scheme for doing so is shown in Figure 5.18. This adjustment is convenient, since it usually does not require any voltage dividers or center-tapped controls. Other circuit-element changes further along the signal path will be ineffective in altering the output zero if there is adequate loop gain. Emitter resistors are not generally added to the input stages for zeroing purposes as they reduce the stage gain. Changes in total collector current are also not recommended as a zeroing procedure, since they do not alter the collector difference of potential. It is possible to unbalance the current in the two transistors to force a balance in base-emitter potentials. Such a procedure can improve the temperature-balance characteristics of the transistors in some cases.

5.16 DIFFERENTIAL OPERATIONAL AMPLIFIERS

A distinction must be made between instrumentation differential amplifiers and differential operational amplifiers. The more significant specifications of an instrumentation device are often obtained by using groups of operational amplifiers. Single gain blocks that use potentiometric and operational feedback techniques in combination to amplify difference signals are referred to as differential operational amplifiers.

The differential input transistors discussed in Section 5-6 are the simplest differential amplifiers. The constant-current source demands that $I_1 + I_2 = I_C$. Difference potentials are converted to amplified potential differences $E_{C1} - E_{C2}$. If both the input terminals are connected to sources of potential, no feedback to these points is possible. This is an open-loop example of a difference (differential) amplifier.

All signals exist as *differences* of potential. The difference can be between a zero reference conductor and a signal conductor or between two signal conductors. Signals referenced to a zero reference conductor are called "single-ended" signals. Conversely, signal differences between two points removed from the zero reference conductor are called a difference or differential signal. This idea is shown in Figure 5.19 in which E_1 and E_2 are voltage differences to the zero reference conductor and E_3 is a difference voltage of interest.

5.17 COMMON-MODE SIGNAL

The potential difference E_0 in Figure 5.8 is independent of the sum of potentials E_1 and E_2. If the difference voltage is zero, then $E_1 = E_2$.

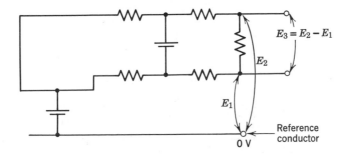

Figure 5.19 Difference versus single-ended signal.

The average value of two voltages is often unimportant, whereas the difference contains information to be amplified. The average signal is called the *common-mode* signal. The desired difference signal could be called the *normal-mode* signal.

This definition of common-mode signal is correct as far as indicated. The accompanying material is not sufficient, however, to define the general problem in instrumentation (see Section 7.7 for more details).

The two signals E_1 and E_2 have an average value, namely, $(E_1 + E_2)/2$. By definition this average voltage or common-mode signal is not amplified by an ideal differential operational amplifier. The next sections describe feedback techniques for amplifying differences and excluding average values.

5.18   OPERATIONAL AMPLIFIER

The basic block of gain used in the previous sections uses a differential input stage followed by single-ended gain. This simply means that a difference signal is amplified independent of average values and presented at an output terminal as a potential difference to the zero reference conductor. The zero reference conductor is often the zero voltage conductor of the power source used to power the amplifier. The open-loop gain configuration is shown in Figure 5.20.

5.19   DIFFERENTIAL AMPLIFIER FEEDBACK

The most commonly used feedback structure around a gain block of Figure 5.20 is shown in Figure 5.21. At balance $E_1' = E_2'$, where E_2' is simply the attenuated value of E_2 or $E_2 R_4/(R_3 + R_4)$ The current I in R_1 is

$$I = \frac{E_1 - E_1'}{R_1} = \frac{E_1 - E_2 R_4/(R_3 + R_4)}{R_1}$$

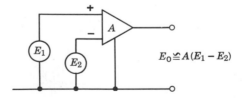

$$E_0 \cong A(E_1 - E_2)$$

Figure 5.20   An open-loop differential amplifier.

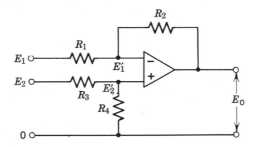

Figure 5.21   A differential amplifier with CMR.

If the input impedance to the amplifier is high, the current in R_1 flows in R_2 and the output voltage $E_O = E_1' - IR_2$. Thus

$$E_O = \frac{E_2 R_4}{R_3 + R_4} - \frac{[E_1 - E_2 R_4/(R_3 + R_4)]R_2}{R_1} \tag{16}$$

For a common-mode signal $E_1 = E_2$ and if the output signal is to be zero, then the balance requirement is simply

$$\frac{R_4}{R_3 + R_4} = \frac{R_1}{R_2}\frac{R_3}{R_3 + R_4} \tag{17}$$

or

$$\frac{R_4}{R_3} = \frac{R_2}{R_1} = G \tag{18}$$

Using this identity reduces (16) to

$$E_O = G(E_2 - E_1)$$

When $E_2 = 0$, (16) reduces to just the operational gain $E_O = -(R_2/R_1)E_1$.

In some applications the two input impedances must be the same; that is, $R_1 = R_3 + R_4$. Since $R_4/R_3 = G$, $R_2 = R_1 G = (R_4/R_3)$ $R_3 + R_4)$.

For applications for which the resistance seen from E_1' and that from E_2' should be equal, then

$$\frac{R_3 R_4}{R_3 + R_4} = \frac{R_1 R_2}{R_1 + R_2}$$

or $R_1 = R_3, R_2 = R_4$.

If the feedback resistor R_2 is returned to an output attenuator, the balance equations derived above are still valid. The loop balance condi-

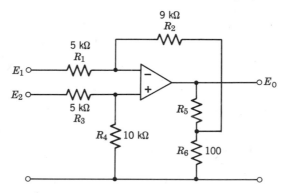

Figure 5.22   An output attenuator applied to Figure 5.21.

tions require the same current flow in R_2, but R_2 must now include
the source impedance of the output attenuator. In the example in Figure
5.22. The parallel combination of R_5 and R_6 must be 1 kΩ so that
when added to R_2, the sum is 10 kΩ $= R_4$. If the attenuator R_5, R_6
is to provide an open-circuit voltage of $E_0/10$ in series with 10 kΩ,
then

$$\frac{R_5 + R_6}{R_1 R_6} = 10 \qquad \text{and} \qquad \frac{R_5 R_6}{R_5 + R_6} = 1 \text{ kΩ}$$

or

$$R_6 = 1.111 \text{ kΩ} \qquad R_5 = 10 \text{ kΩ}$$

Since $G = 10 \text{ kΩ}/5 \text{ kΩ} = 2$, the total gain is $G \times 2 = 20$.

In Figure 5.22 variable-gain controls are difficult to apply without
upsetting the balance of resistor ratios. Of course, if R_2 is small enough,
varying R_1 will not seriously affect the common-mode rejection (CMR)
ratio. For large gain values R_2 is small and varying R_2 is acceptable.

5.20   CONSTANT-CURRENT SOURCES

Some feedback configurations raise the output impedance. An infinite
output impedance is synonymous with a constant-current source. An
infinite output impedance implies that the output voltage is proportional
to load value; that is, E/R is a constant. A test for a constant-current
source is to halve the load resistance and observe whether the signal
voltage also halves. When the test is performed, the measuring device
must also be considered as part of the load R.

A simple feedback structure with constant-current output is given in Figure 5.23. At balance the potential $E_{IN} - E_2$ is near zero, or $E_{IN} = E_2$. The current in R is thus $I = E_2/R = E_{IN}/R$. This current must arise from the amplifier output via R_L. The output voltage is $E_0 = IR_L = (E_{IN}/R)R_L$. Note that E_0 is proportional to R_L or an infinite output impedance.

Another way of viewing the output impedance is from the definition of Z_0:

$$Z_0 = \frac{\Delta E}{\Delta I}$$

As the load changes, ΔE changes but ΔI does not change. Division by zero results in Z_0 being infinite.

The input impedance of the amplifier in Figure 5.23 is increased by the feedback factor over its open-loop value. Since $E_{IN} = E_2$ plus an error signal, signal current flow for the potential difference $E_{IN} - E_2$ is very small. This impedance concept includes only signal currents, not static values or "pumpout" values. If the "pumpout" current is provided from the power source, the input impedance is still high as this resistance is removed from its shunting effect by the feedback factor.

The power supply in Figure 5.23 is not referenced to the zero of signal potential. Supplying power to such an amplifier is difficult and requires special transformer shielding. The proper shield treatment is shown in Figure 5.24. Note that a leakage capacitance C_{12} permits transformer current to flow in R via the loop ① ② ③ ④ ①. This current is treated as signal and is amplified. No amount of power-supply filtering can reduce this unwanted pickup.

A feedback configuration with constant-current output but with a nonfloating power source is shown in Figure 5.25. The voltage E_0 also appears at E_1. The current flowing to signal zero from the output

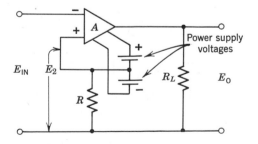

Figure 5.23   A constant-current source.

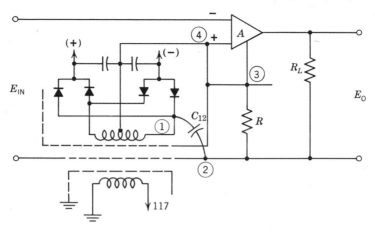

Figure 5.24 Shielding a transformer for a constant-current source.

terminal is $E_O/R_L + E_O/R_3$. It flows from the amplifier output through R_4. Thus the amplifier output is

$$E_A = E_O + IR_4 = E_O + \frac{E_O R_4}{R_L} + \frac{E_O R_4}{R_3}$$

Since $E_2 = E_1 = E_O$, the current into R_1 from the source is $(E_{\text{IN}} - E_O)/R_1$. Thus E_A is just the potential at E_2 minus the potential drop across R_2, or

$$E_A = E_O - (E_{\text{IN}} - E_O)\frac{R_2}{R_1} = E_O\left(1 + \frac{R_4}{R_L} + \frac{R_4}{R_3}\right)$$

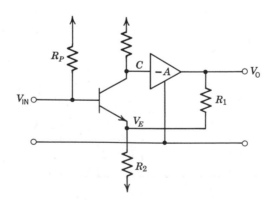

Figure 5.25 A current source using a grounded power supply.

Solving for E_O yields

$$E_O = \frac{R_2}{R_1} E_{\text{IN}} \left(\frac{R_4}{R_L} + \frac{R_4}{R_3} - \frac{R_2}{R_1} \right)^{-1}$$

If E_O is to be proportional to R_L, the only condition to be met is that

$$\frac{R_4}{R_3} = \frac{R_2}{R_1} \tag{19}$$

By using this relationship the output voltage can be written as

$$E_O = E_{\text{IN}} \frac{R_L}{R_3} \tag{20}$$

This circuit thus has infinite output impedance if condition (19) is met. The current gain is completely determined by the resistor value R_3. If E_4 is to be kept within bounds, R_4 should be kept less than the parallel combination of R_L and R_3. (Note this circuit does not have the high input impedance of the circuit in Figure 5.24.)

5.21   INSTRUMENTATION DIFFERENTIAL AMPLIFIERS

The performance of IC operational amplifiers configured as differential amplifiers is limited. The common-mode levels and rejection ratios are inadequate for many practical system problems. The output current and the high-frequency performance are often inadequate. Offset and gain adjustment ranges must be handled externally.

For these reasons combinations of IC amplifiers or discrete circuit techniques in combination with IC amplifiers are used to satisfy many user requirements. This class of amplifier is called an instrumentation amplifier. To make the full discussion meaningful, some discussion of discrete-differential-amplifier technique is useful.

5.22   EMITTER FEEDBACK

Operational amplifiers of the IC type are usually limited to two input connections and an output connection. Pins are provided for stabilization networks and for operating power. Other internal circuit points (emitters and sources) are not generally brought out for external use. In discrete-component amplifiers feedback can be applied to any desired node for improved circuit performance. Since this feedback technique is important, the details are discussed below.

The simplest active feedback structure involves a single first-stage transistor. Signals returned to the emitter after suitable gain constitute

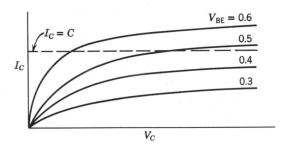

Figure 5.26 A simple active feedback structure.

a class of potentiometric feedback. If the base moves in potential, the emitter is moved to maintain the same base-emitter potential difference. When the static input current is supplied from a secondary source, the signal current is the only current that flows in the input lead. With feedback this signal current is very small, and thus the input impedance is high.

For a differential input pair of transistors the feedback base drives the input emitter. This is exactly the circuit described above with the addition of a driving or feedback transistor. The emitter can be considered a low-impedance source providing isolation for the signal applied to the feedback base.

When feedback connections are made directly to an emitter, signal current and operating current share the same resistor. Since the transistor is in the feedback loop, its characteristics are a part of the resulting transfer characteristic. Consider the circuit in Figure 5.26. If A is a large

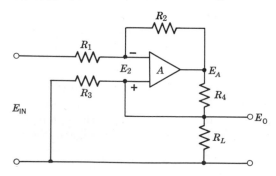

Figure 5.27 A typical operating family of I_cV_c curves for various V_{BC}.

value, the signal at C is small. Thus as V_{IN} changes, V_E changes to keep point C constant. As a first approximation $V_{IN} = V_E$ and $V_0 \cong V_{IN} (R_1 + R_2)/R_2$. On closer examination, as V_{IN} increases, the collector voltage is reduced; V_{BE} must therefore assume a difference that keeps the current fixed to keep point C constant. Figure 5.27 shows a typical transistor curve. If I_C is held constant, V_{BE} varies as a function of V_C. If $\Delta V_{BE}/\Delta V_C$ is linear, the linearity will reflect in a linear gain-correction multiplier (a further factor in the β loop). If $\Delta V_{BE}/\Delta V_C$ is nonlinear, the nonlinearity will appear in E_0 independent of the value of A.

5.23 MULTIPLE-FEEDBACK LOOPS FOR COMMON-MODE REJECTION

If two transistors are used with separate emitter resistors, feedback signals can be brought to both emitters in a differential sense. A circuit with such an arrangement is shown in Figure 5.28. Structures using the multiple-feedback technique can be the basis of an instrumentation-type differential amplifier. The point CM between R_3 and R_3' is a common-mode feedback point. A current injected at point CM causes nearly equal currents to flow in R_2 and R_2'. As a first approximation, this current does *not* add to the difference signal appearing between V_{C_1} and V_{C_2}.

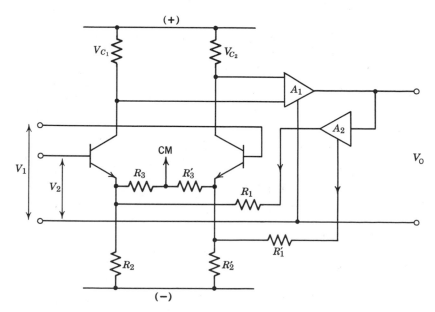

Figure 5.28 Differential feedback scheme with separate emitter resistors.

Current flowing in R_1 and R_1', however, constitutes differential feedback; that is, a *difference* current flowing in R_1 and R_1' has a *difference* transfer gain to the collectors V_{C_1} and V_{C_2}. The point CM can be connected to the output of an amplifier that has gain to the average signal $(V_{C_1} + V_{C_2})/2$. If the loop gain to the common-mode signal is high, the average signal will be reduced to near zero. The feedback ensures that an input common-mode signal $(V_1 + V_2)/2$ causes a small error signal, namely, $(V_{C_1} + V_{C_2})/2$. This is identical in effect to the constant-current source used in previously mentioned circuits in which the emitters are tied together. In this case feedback is used to ensure a single average value of current, whereas in the previous case a constant-current source did the same thing.

A block diagram showing normal-mode and common-mode feedback is shown in Figure 5.29. An analysis of circuit performance can be made by considering normal-mode and common-mode processes separately. If $V_1 = V_2$, the input signal is entirely common mode. If the transistors are assumed balanced, the emitters follow the signal potentials except for a fixed base-emitter potential. In this analysis, only signal differences are of concern, and the emitter signal potential thus equals the base

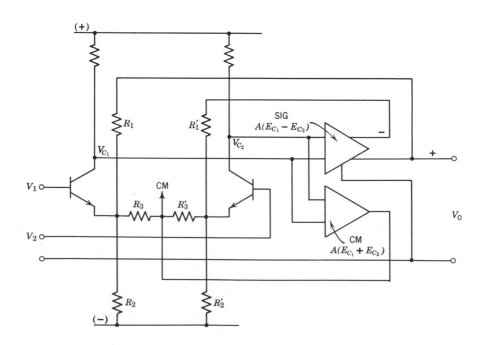

Figure 5.29 Normal-mode and common-mode feedback.

signal potential. Since the average value $(V_{c1} + V_{c2})/2$ remains essentially constant, the output of the common-mode amplifier supplies changes of current to R_2 and R_2' as demanded by changes in emitter potential. If $R_3 = R_2$, the signal output voltage of A_{CM} is twice the input signal potential V_1.

If signal voltages V_1 and V_2 are equal and opposite in sign, the signal has no average or common-mode component. Since the emitters move equally and oppositely, the point CM remains at zero signal volts. The output voltage V_0 provides an emitter potential equal to the input signal V_1. Since the point CM is at zero, R_3 loads the emitter in parallel with R_2. The gain is determined by the ratio of R_1 to the parallel resistance of R_2 and R_3; that is,

$$G = \frac{R_1 + R_{23}}{R_{23}} \quad \text{if} \quad \frac{R_2 R_3}{R_2 + R_3} = R_{23}$$

The common-mode amplifier A_{CM} could be a voltage or a current source, and the voltage at point CM is not altered. The point CM is at zero potential for normal signals independent of the presence of the common-mode amplifier. The common-mode circuit becomes inoperative if $R_1 = R_1' = 0$ at a gain of 1. At this gain the low output impedance of the signal amplifier is a short circuit as viewed from the common-mode amplifier, and the common-mode loop is open.

At low gains, large signal swings are required of V_1 and V_2. Resistors R_1 and R_1' are low in value, and large signal levels from A_{CM} are required. Since the intent of A_{CM} is to take up the current needs of the input stages, A_{CM} can be replaced by two equal current sources connected directly to the emitters. Since the emitter signals never exceed the input signals, this is also the voltage limit for the current sources. (The voltage swing of a current source is called its compliance.)

Common-mode signals are usually large signals in the volt range, whereas normal signals are perhaps only millivolts. If an input stage is to process both signal types simultaneously without overloading, the dynamic qualities of the input stage are severely tested. Low drift and noise might demand a low collector current, while common-mode voltage swing may dictate a larger current. The simultaneous high-frequency stabilization of both normal-mode and common-mode processes may demand stabilization networks not compatible with the current requirements of high-frequency common-mode signals. When a proper compromise cannot be found, other techniques must be considered.

The currents flowing in the capacitances of the input stages can be reduced if the power source for the input stage is driven by the common-mode signal. Such a procedure assures that the base-collector capacitance

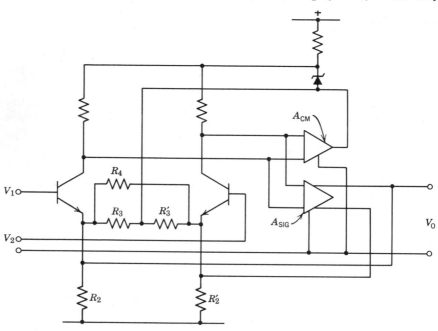

Figure 5.30 Power-supply feedback for CMR.

is reduced. It raises the common-mode loop gain and improves the frequency characteristics of the normal-mode amplifier (see Figure 5.30).

5.24 GAIN CHANGING AND OFFSET

Multiple resistor switching is necessary if feedback resistors R_1 and R_1' are switched. (R_2 and R_2' determine collector currents and are not available for switching). Both switched resistors must be equal to maintain input balance. Resistors R_3 and R_3' can be switched to change gain, but these resistors also change the gain in the common-mode feedback loop. A single bridging resistor R_4 between the two emitters as shown in Figure 5.30 is a preferred solution to the gain switching as this resistor does not affect the common-mode loop gain. Resistor R_4 can also serve the purpose of providing continuously variable gain. The equations for gain given earlier provide the basis for a calculation. The resistor R_4 cannot go to 0 Ω or the signal loop will be opened. Note that the common-mode signal current does not flow in R_4 and thus its presence does not affect the output signal level of A_{CM}.

The amplifier shown in Figure 5.30 can be offset RTI by adding current to either emitter from an external source. This current has both a normal signal and a common-mode signal component.

PROBLEMS

1. Design a variable-gain circuit as in Figure 5.1 varying R_{FB}. If $R_1 = 100$ kΩ, design the two gain extremes to be 100 and 200.

2. In Figure 5.3 assume no attenuator loading. Design a variable-gain circuit to span the gains 100 to 333.3.

3. If the amplifier in Figure 5.5 has an $A\beta$ of 40 dB, how much signal current will flow in R_{IN}?

4. In Figure 5.5, if R_2 is 10 MΩ, how much noise voltage will the resistor produce in the output in 10-Hz bandwidth?

5. The input impedance of an amplifier is 1000 MΩ. What input cable capacitance reduces the input impedance to 707 MΩ at 100 Hz?

6. Using Figure 5.10 and Equation 15, design a variable-gain control to cover the gain range 1 to 11 (sensitivity of 1 and 0.091). Use a 10-turn control that reads sensitivity correctly at counting dial settings of 1 and 10.

7. Show that an input attenuator reduces RTI specifications by the attenuation factor. Use noise as a typical specification.

8. In Figure 5.15 prove that $E_0 = E_s R_3 / R_5$ independent of R_2

9. The gain of an amplifier is given by $R_1 / (R_1 + R_2)$ where $R_1 = 10$ kΩ, $R_2 = 1$ kΩ. Input current flows in R_1 and R_1 is varied from 10 to 11 kΩ. If the output voltage changes from 10 to 20 mV, what is the value of the input current?

10. A 1-mV signal is impressed on a differential amplifier with one input grounded. What is the normal-mode and common-mode signal?

11. Use Figure 5.21 as a model and design a gain 5 amplifier. What is the purpose of R_4 if E_2 is zero.

12. In Figure 5.21 indicate the signal levels at the amplifier's input terminals if $E_1 = +1$ V and $E_2 = -1$ V. If $E_1 = +2$ V and $E_2 = 0$ V.

13. In Figure 5.21 what is the differential input impedance?

14. If $E_1 = 0$ and $E_2 = 1$ V, how much current flows in both input resistors?

15. An output voltage is 1.0 V with a 1-kΩ load and rises to 1.8 V with a 2-kΩ load. The meter is a 100-kΩ load. What is the output impedance of the amplifier?

16. In Figure 5.30 explain why common-mode current does not flow in R_4.

6

Modulation and Demodulation

6.1 INTRODUCTION

Modulation and demodulation processes are in effect phase-sensitive rectifications. The specific methods used to convert a dc signal to an ac signal and return can vary significantly. If the basic ideas are clear, the details of a design can easily be grasped.

6.2 HALF-WAVE MODULATION

Consider a dc signal and a switch S as in Figure 6.1a. When S is closed, $E_0 = 0$; when S is open, $E_0 = E_B$. If S operated 400 times a second, the signal waveform for E_0 is that shown in Figure 6.1b or c and is a 400-Hz square wave. Figure 6.1b represents the waveform if S is open 50% of the time. In Figure 6.1c S is open 80% of the time. The switch S has converted a dc signal to ac.

If an output coupling capacitor is used as in Figure 6.2a, the average dc value is removed, leaving waveforms as shown in Figure 6.2b and c. The resistor R limits the current flowing in the switch. The peak-to-peak signal output is always just E_B. This process is called *half-wave modulation*. If a load resistor is applied to the output terminals, the size of capacitor C determines the sag phenomenon associated with each cycle of the output signal. The load resistor also attenuates the output signal by just the ratio $k = R_L/(R_L + R)$. Waveforms for the loaded

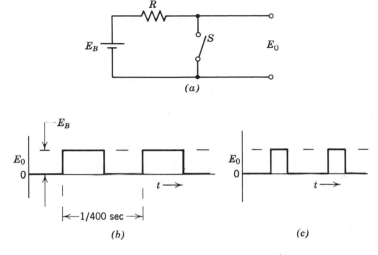

Figure 6.1   A simple modulator and resultant waveforms.

capacitor-coupled case are shown in Figure 6.3. If capacitor C is small, the sag phenomenon is increased. Signals with excessive sag are inefficient and are usually undesirable.

The ac signal output in Figure 6.3 is proportional to the signal E_B.

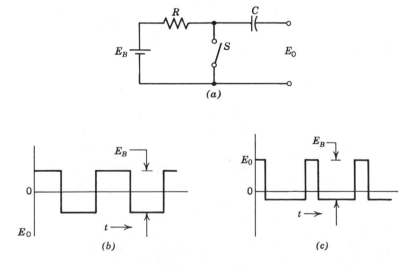

Figure 6.2   A modulator and an added coupling capacitor.

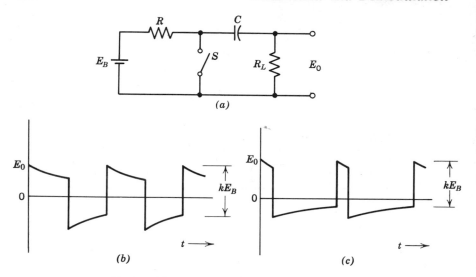

Figure 6.3 A modulator with RC coupling.

If E_B is reversed in polarity, the output signal inverts as shown in Figure 6.4. At the moment the switch opens, the output signals in Figure 6.4b and c go negative; that is, the ac signal has a timing relationship (phase) that is a function of the input signal polarity.

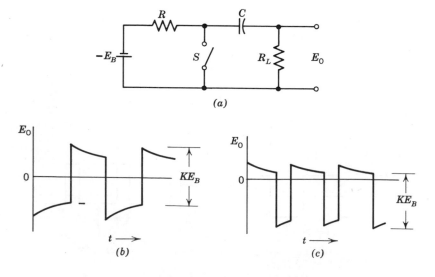

Figure 6.4 Modulation with a negative input signal.

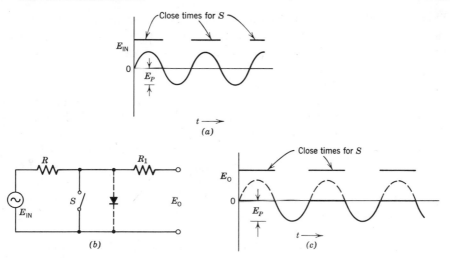

Figure 6.5 A synchronous switch applied to a sine-wave signal.

6.3 DEMODULATION

The switch S may also act as a demodulator; that is, it can transform an ac signal to dc. If the signal in Figure 6.5 is a 400-Hz sine wave, the switch S can be operated 400 times a second to form a half-wave rectifier. Consider that switch S is closed during positive halves of this input sine wave. The resulting output waveform is shown in Figure 6.5c.

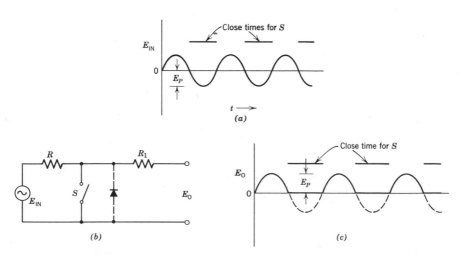

Figure 6.6 A synchronous switch applied to a sine-wave signal. Note switch closures are 90° shifted from Figure 6.5.

This is exactly equivalent to the switch S being replaced by an ideal diode, shown dotted. When the open and close times are interchanged on the switch, the output signal appears as in Figure 6.6. Note that now the output signal becomes a series of positive-going half sine waves, equivalent to inverting the dotted diode in Figure 6.5. This process is called *half-wave demodulation*. The timing of switch closures in Figures 6.5 and 6.6 controls the polarity of rectification. When the switch transitions occur at peak values of the incoming sine wave, the resultant waveforms are those shown in Figure 6.7. The average value of the two output signals is zero. It should be apparent that the average value of the output signal is maximum when the switch operations are timed to coincide with zero crossings of the input sine wave. In Figure 6.7 the output signal is symmetrical, plus and minus, during the switch-open period, and this results in no net dc output.

The half-wave waveforms in Figures 6.5 and 6.6 can be smoothed by adding a filter capacitor C_1. The resulting output signal is shown in Figure 6.8c. If R_L is large, the time constant $R_L C_1$ determines the sag per cycle of the filtered output. The value of $R_L C_1$ thus controls the peak-to-peak output ripple and can be made large enough to reduce the ripple to acceptable levels. If R_L loads the source, then the attenuating effects of R and R_1 must also be included. On the contact half cycle, C_1 is discharged through the parallel combination of R_1 and R_L, and, on the open half cycle, C_1 is charged through the series combination R and R_1 and discharged through R_L. The algebra for treating the general case is messy and contributes little to the general understanding. The problem is even

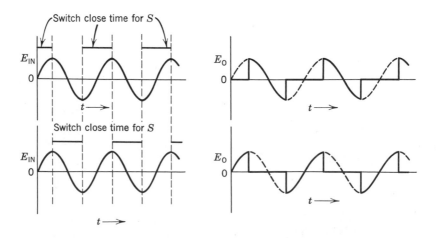

Figure 6.7 Waveforms for switch closures at peak values.

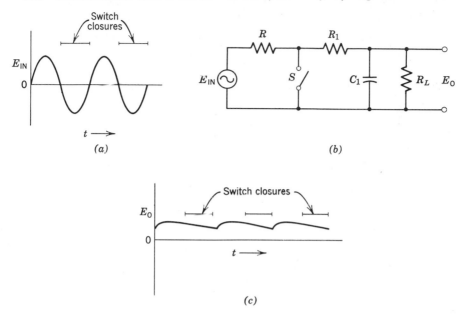

Figure 6.8 Smoothed or filtered output signals from a half-wave demodulator.

more complex if the switch contact and open times are not equal or symmetrical.

Useful demodulation can occur for unsymmetrical switch open and close times. (The ratio of close to open time is called the contact duty cycle.) This operation has lower efficiency because a part of the signal that could have been rectified is ignored. If sine waves are being demodulated, the dominant content occurs near peak values and the signal loss near the zero crossings is small. If square waves are being demodulated, the percentage loss of signal is linearly related to the departure of contact dwell time from the ideal. A typical demodulated square-wave signal resulting from unequal duty cycle is shown in Figure 6.9.

6.4 MODULATION AND DEMODULATION (MODEM) OF THE SAME SIGNAL

When a dc signal is modulated and amplified, the resulting ac signal can be demodulated and filtered to produce a new dc signal. Because of the ac amplifier gain, *a net dc gain can result*. The waveforms of a typical half-wave modulator and demodulator are shown in Figure 6.10*a* and *b*. This scheme is properly called a dc amplifier. The switch

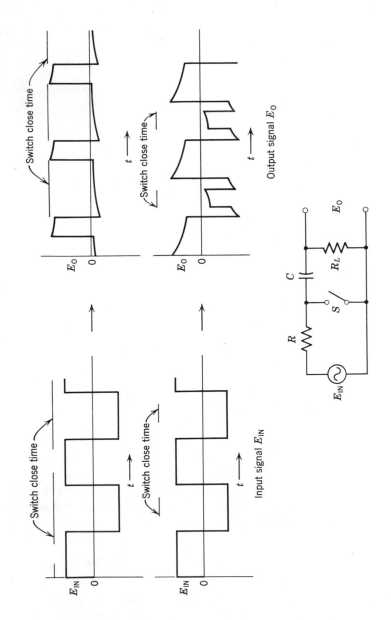

144

Figure 6.9 Unequal dwell times and their waveforms for square-wave input signals.

closures in (*a*) occur together. The first switch produces an ac signal voltage proportional to the input signal. The second switch rectifies the ac output signal voltage. The output filter R_3C_3 smoothes the result. The net dc gain is related to contact efficiencies and the gain in the ac amplifier.

Two variations in this circuit are possible. First, the ac amplifier can have a negative gain; that is, the output goes positive for a negative-going input. Second, the switch closure times on the demodulator can alternate (can be shifted one half-cycle in time) with the input modulator. When the ac amplifier is negative gain, the resultant dc gain is negative. When the demodulator switch closure is shifted one half-cycle in time, the dc gain also reverses. If the amplifier gain is negative and the demodulator is shifted one half-cycle, the resultant dc gain is again positive. Waveforms for this case are shown in Figure 6.10*b*. Note that resistor R_2 limits the demodulator switch current drawn from the amplifier output stage and capacitor C_2 ensures that any dc component in the amplifier output stage will not add to the expected rectified signal value.

The waveforms in Figure 6.10 are steady state. They do not include changes in input signal or low-frequency transient processes in the ac amplifier. Before any modulation occurs, the charge on the coupling

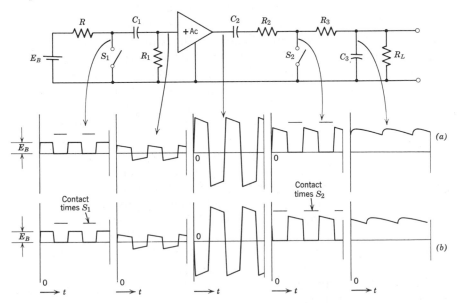

Figure 6.10 Waveforms in a half-wave modem.

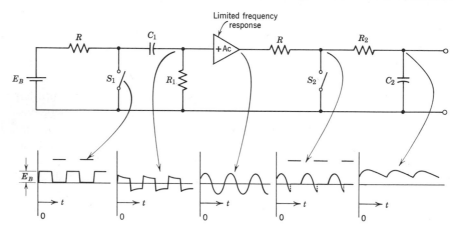

Figure 6.11   Waveforms in which amplifier bandwidth is limited.

capacitors is zero. In the steady-state condition this average charge is
not zero, and a number of cycles of modulation must pass before a
nontransient condition exists.

The ac amplifier in Figure 6.10 is wide band; that is, the output
signal faithfully matches the input square wave. In many applications
the bandwidth of the ac amplifier is limited. When frequencies above
the modulation frequency are attenuated, the leading edge of the incoming square wave is rounded. If the low-frequency response is attenuated,
the sag phenomenon is increased. If the amplifier response is "peaked"
in the vicinity of the modulation frequency, the output waveform can
appear somewhat sinusoidal. This type of waveform can still be demodulated provided the zero-axis crossings are timed to relate to the switch
closure periods.

If the modulated signal is being amplified as a sinusoid, the phase
of the output amplified signal is important. The demodulator switch
closures and the output signal must be phased together for the demodulation to be effective. When the amplifier attenuates only high frequencies,
midband signals are delayed in phase. When only low frequencies are
attenuated, the midband signals lead in phase. When the frequency-
versus-amplitude response is peaked symmetrically about the modulation
frequency, the phase shift at the modulation frequency is zero. Zero
phase shift is usually sought to optimize modulation-demodulation processes. This is the timing mentioned above.

Figure 6.11 illustrates the waveforms that can result when an amplifier
has reduced high-frequency bandwidth. The slight phase shift that results

distorts the output signal waveform. It is desirable to have the demodulator switch closure occur at a zero signal crossing, which is possible only when the modulation signals are specially timed and shaped (sinusoidal) or when the content of the modulation signal has been filtered to produce a smooth signal transition through zero voltage. This treatment is particularly important when the demodulated signal is brought physically near the input stages of low-level amplifiers. The transient phenomenon associated with switch closures is easily coupled directly into critical input circuits.

6.5   FULL-WAVE MODULATION AND DEMODULATION

A full-wave modulator, as the name implies, uses the signal almost full time instead of shorting out the signal path on successive half-cycles. Full-wave processes can use a transformer as in Figure 6.12. When the reed of the switch is "up," the input signal E_B is connected between the center tap and the top lead of the transformer. By transformer action, an equal and opposite signal appears across the bottom half coil of the transformer. When the reed of the switch is "down," the input signal appears between the center tap and the bottom lead of the transformer. The voltage across the bottom half is equal to the signal source voltage and equal and opposite in polarity to the induced signal of the preceding half-cycle. The peak-to-peak signal appearing between the center tap and each lead of the transformer is double the input signal. The peak-to-peak signal across the entire primary coil is four times the input signal.

The demodulation of an ac signal can take on any standard rectifier form. Since rectification can be full-wave center tapped or voltage doubling, so can demodulation. Switch closure times correspond to forward conduction times for diodes. Since the closures can be shifted 180°, the diodes can in effect be reversed. When diodes are reversed in normal

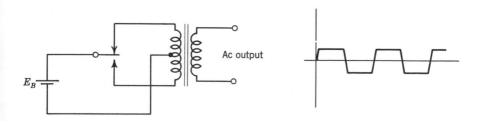

Figure 6.12 A transformer-coupled full-wave modulator.

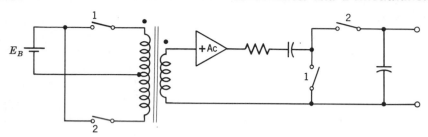

Figure 6.13 A full-wave modulator and voltage-doubler demodulator.

rectification, the output polarity reverses. When demodulation switch closures are shifted 180°, the output polarity also reverses. Figure 6.13 shows a full-wave voltage modulator and a voltage-doubling demodulator. The similarly numbered switches are closed at the same time. The circuit in Figure 6.13 has a negative gain at zero frequency. If the gain of A is negative, the dc gain is positive.

6.6 PRACTICAL MECHANICAL MODULATORS FOR LOW-LEVEL APPLICATIONS

The contact closures in the preceding discussion could be magnetically operated. Such an arrangement of contacts is often called a "chopper." Here contacts are made of precious metal and care is taken to shield the contact driving function from the signal function.

In low-level chopper applications many difficulties arise. The heat of the drive coil can flow to the contacts causing thermocouple action. It is impossible to remove the heat flow, but if the mass of the switch is adequate and the mechanical design is symmetrical the emfs that are developed can be effectively balanced out. The problem of symmetry obviously carries over to the modulator base contacts and even to the wiring that engages the base. The circuitry that connects to the modulator is usually not symmetrical, and the only remedy is to keep the base and this circuitry away from any external sources of heat.

The problem of symmetry and thermal emfs is straightforward. Any junction of dissimilar metals at a temperature removed from points of observation develops an emf. If equal and opposite junctions are placed in the circuit at the same temperature, the emfs cancel. The principal difficulty lies in ensuring equal temperatures at paired junctions, not in providing the junctions. The trouble that usually occurs is that the circuits connected to the contacts do not have equal thermal capacity or equal radiation characteristics.

One procedure used in very sensitive low-level applications involves special low-thermal solder. Its use reduces the effects of conductor-to-solder emfs but does not reduce the emfs that occur between two dissimilar bonded conductors. It is still advised that the most attention be paid to heat flow and temperature equivalence at points of symmetry.

The magnetic field from the drive coil of a chopper can couple to the contact circuit. Since the contacts and the input circuitry must constitute a physical loop, the chopper manufacturer pays a great deal of attention to keeping the external field in the vicinity of the base pins low. When this field is sensed in the input leads, it cannot be differentiated from signals developed by modulating an input signal. The error may be quite stable, but it is an undesirable effect to have in a design. If the drive voltage changes in either waveform or level, a change in input signal results. If any change in external magnetic geometry occurs, an input signal results. If the chopper is physically moved or is vibrated to a new position, a new input signal can result.

In using a chopper, an optimum geometry for input lead dress can be found. This dress reduces the magnetic field sensed in the input loop and also ensures a symmetrical heat flow away from the chopper contacts. This area is sensitive in low-level designs and is the reason for one design being better than another, using the same components and hardware.

Contact closures do not occur at the same time on each drive cycle. This uncertainty is often called contact bounce. In some designs, the contacts are "slammed" across so that the contact time is essentially independent of drive voltage. This type of drive produces a larger magnetic field and often yields a better chopper. The effect of contact bounce varies depending on how the chopper is used. When the chopper is used at null, that is, when the signals are always zero, the effects of bounce are minimal as a change in dwell time produces no transient signal. If the chopper is operating at null but undesired higher harmonic signals are present, the effect of bounce can be significant. These higher harmonic signals can be picked up from the chopper drive coil or from external sources if they are coupled parasitically at the points of contact.

The "bounce effect" in the presence of harmonic pickup is dependent on the coupling mechanism involved. If second-harmonic content is present in a half-wave modulator (see Figure 6-10), the input coupling capacitor stores an average charge dependent on the signal level. If bounce occurs, the charge can shift in value and the shift appears as transient output. The transient would not have occurred if the undesired content had not been present. This is further reason for keeping the parasitic input coupling very low rather than relying on some canceling or balancing scheme.

The magnetic field near the contact point after contact closure changes abruptly as the flux captured by the contact loop must dissipate energy in that loop. The flux collapse is usually rather sudden and can result in "spike-type" signals in the amplifier input. Again the effect can be minimized by keeping the magnetic fields away from the contact circuit loops and keeping loop areas minimum.

The contacts themselves are subject for a great deal of discussion. This area of engineering (or black magic) is best left to the chopper manufacturers. The designer needs only to understand the difficulties in evaluation and application, which are a significant problem.

In the modulation scheme of Figure 6.10*b* the switch closures are shown separately. In practice the two switch closures can be made by the same switch element as a single-pole double-throw switch. This procedure is confusing to some because of the way the schematics are drawn. The switches are independent however and operate as shown. When a single reed is used, the proximity of input circuitry to output signal points can cause design problems, particularly if large ac gains are used. The presence of just 1 pF of coupling capacitance can greatly modify the performance of the ac amplifier. Here the circuit performs in a manner very dependent on contact duty cycle. For large ac gains no time should exist when the input or the output is not shorted out. If the input can "see" the output, an oscillating mode can start each cycle. The amplifier can be unstable when the modulator drive is disconnected or when the chopper is removed from its base. The troubleshooter should be aware of this possibility.

6.7   MODULATION/DEMODULATION DEVICES—GENERAL

The switch contacts used in the previous sections provide a near ideal modulator. The short-circuit resistance is almost 0 Ω, and the open-circuit resistance is essentially infinite. The ratio of the short-to-open circuit resistance is one measure of modulator efficiency. It is not necessary for the ratio to be high for the modulator to be effective as the modulation process needs only to produce an adequate ac signal.

The essential quality of a modulator element is its nonlinearity. When the nonlinear quality is controlled, modulation can result. Most modulators can be characterized as variable resistors with the resistance changed by some driving signal. A FET transistor or a photosensitive semiconductor can be used to modulate a signal. When the gate of the FET is biased on, the FET junction can appear as a few hundred ohms, and when biased off, the resistance can jump to many megohms. This change in resistance can be used to modulate a signal. Such a circuit is shown in Figure 6.14. When a depletion-mode FET is used and the input square wave is at 0 V, the FET switch is conductive. A signal loss occurs because of the resistive divider $R_L/(R_L + R_s + R_{IN})$. When the square-wave input voltage is negative, some signal still appears at the input, namely, $R_L/(R_0 + R_{IN} + R_L)$. The ratio of signal magnitudes, with R_0 and R_s set to practical and then ideal values, defines the effectiveness of the modulator.

The modulator of Figure 6.14 is sensitive to any coil drive signal that couples directly into the amplifier input. The coupling is dependent on R_{IN} and the capacitance between the gate and the FET's conductive path. The static currents that flow to the gate can be held to below 1 nA. Careful construction and diode clamping can reduce the undesired dynamic coupling to near this same value.

The FET element in Figure 6-14 can be replaced by a photosensitive element. A change in resistance can be effected by shining light on the element. As the light is varied, an ac signal can be generated. Such a modulator is not subject to the drive-coupling difficulties of a FET switch, but the heat developed by the light must be carefully considered.

Many elements can serve as a modulator. They might include the following:

1. A mechanically stressed capacitor. When a charge Q is placed on a capacitor, $\Delta E = -(Q/C^2)\,\Delta C$.

2. A variable reluctance path. Here a mechanical motion can unbalance the difference between two reactive signals.

3. A strain gage. A resistance bridge can be unbalanced by mechanical strain.

Demodulators usually function on high-level signals, and, because of this, they are usually not subject to the same parasitic difficulty as low-level modulators. The process of rectification is the same however; an element is caused to operate in a nonlinear manner to rectify an

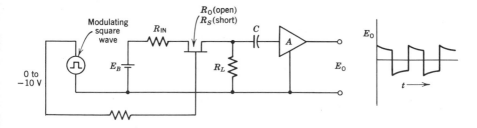

Figure 6.14   A FET modulator.

ac signal in a phase-sensitive manner. Many practical high-level demodulators (modulators too) are just transistors turned to a full-conductive
state for low resistance and turned off for the nonconductive state.
"Turn-on" base current flows during the "on state" when the transistor
resistance is low. This state can correspond to a shunt connection to
the zero reference conductor. In this way the zero-error effects of base-
current flow can be minimal.

Two transistors connected in series as in Figure 6.15 can be used
as a modulation element. The emfs across the two switching transistors
are nearly equal. If the transistors are carefully matched, the net emf
can be held to tens of microvolts. Heating of the transistors from the
modulation drive causes the balance to shift. If large signals are modulated, the back voltage applied to the switch on the "open" half-cycle
can also add to the heating. Even if the transistors are matched over
a useful temperature range, unequal heating will nullify any effort spent
in matching.

The paired-transistor switch element of Figure 6.15 is often used for
high-level demodulation. Here even a 1-mV base-emitter difference must
be considered as it is 0.1% of a 1-V signal. Transistors are selected
for their low "on" resistance or for their ability to handle high reverse
voltages depending on application. Note that if the input voltage in
Figure 6.15 is positive, transistor A "holds off" the input voltage. If
the voltage is negative, transistor B "holds off" the voltage and transistor
A is forward conducting.

The modulation drive signal in Figure 6.15 is isolated ohmically from
the signal common by a transformer. This procedure is not mandatory,
but it is often necessary in demodulation schemes when signal changes
relative to signal common occur during the "on" or low-resistance portion

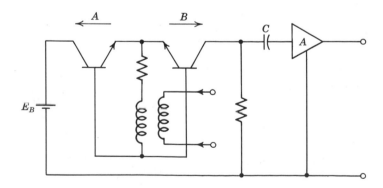

Figure 6.15 A transistor pair as a switch element.

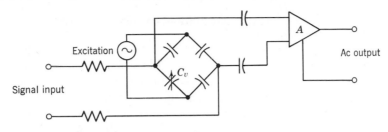

Figure 6.16 A capacitor modulator.

of the switching cycle. In this specific circuit, "on" current flows in both transistor bases at the same time. On the opposite half-cycle, the base-emitter diodes are reverse biased and little current flows. The result is a net dc ampere-turn flow in the transformer. If a second switch pair is involved elsewhere in the circuit that is driven from the same transformer, the net dc ampere-turn flow can be held to near zero and no core saturation will result.

A modulation scheme sometimes used in electrometer amplifiers involves an input capacitor bridge. One or two of the capacitors have a voltage coefficient of capacitance so that the bridge is unbalanced in the presence of an input signal. The capacitor bridge is excited by an ac signal, and the unbalance develops an ac signal for amplification. In effect a nonlinear element modulates an input signal. The result is an ac signal proportional to the input signal. Feedback voltage can be used to return the bridge to null. An open-ended circuit with one variable capacitor C_V is shown in Figure 6.16.

In low-level optical detectors the incident light energy is often mechanically "chopped" by a moving mechanical blade operated from 60-Hz power. The resultant ac signal from the detector is amplified and demodulated by a signal that is synchronous with the power frequency. Because 60-Hz pickup can be a problem, the chopper frequency is often set at 30 Hz to avoid the difficulty. Generation of a 30-Hz demodulation signal is not difficult.

When strain-gage bridges are ac excited (100 to 10,000 Hz), a gage strain can produce a proportional signal at the excitation frequency which can then be amplified. The strain in effect acts as a modulator. The same excitation signal used to operate the bridge can be used to operate the demodulator which converts the amplified ac signal back to strain information. In such systems, attention must be paid to obtaining reactive bridge balance as well as resistive balance.

6.8   CARRIER-SUPPRESSED MODULATION

A carrier system, as the name implies, carries low-frequency information on a higher-frequency signal. In radio transmission, the information is contained either as amplitude or frequency modulation. In voice communication systems, the modulating signals are large, and the zero reference amplitude or frequency is not used to convey information. The carrier levels used in instrumentation carry dc information. For this reason much attention needs to be paid to static or reference levels.

The modulation schemes shown in Sections 6.4 and 6.5 have one feature in common; namely, if there is no input signal, there is no resultant ac signal. Such modulation is known as carrier-suppressed modulation. The phase of the carrier signal relative to the drive signal of the modulator contains the signal polarity information. The amplitude of the resultant carrier signal is proportional to the level of the signal being modulated. All of the preceding examples had a battery as a signal source. If the source is varied in a sinusoidal manner, the resultant modulation varies in magnitude in a sinusoidal manner.

If the signal being modulated is at a low frequency compared with the modulation frequency, a Lissajous pattern of output signal versus input signal is a butterfly pattern as shown in Figure 6.17. The pattern

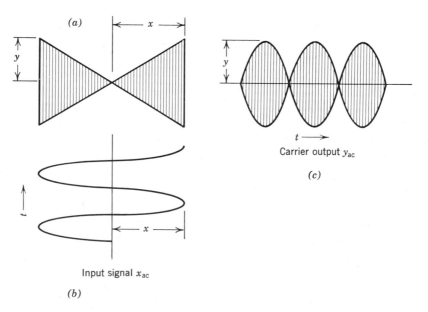

Figure 6.17 A Lissajous pattern of carrier-suppressed modulation.

is the same if the signal being modulated is triangular or nonlinear in shape. The ac signal level output y_{ac} is proportional to the input signal x independent of waveform.[1] Mathematically

$$y_{ac} = kx$$

The butterfly pattern becomes complex when there is any time delay between the x- and y-axis signals or when the two frequencies approach each other. Patterns such as that in Figure 6.17 require that the oscilloscope observation be made with the oscilloscope set for a dc response on both the x and y axes.

The process of modulation places the signal information onto the carrier. This figurative description can be supported by a more mathematical treatment. For obvious reasons it is simpler to consider only sinusoids. If a modulated carrier signal ω_c is proportional to an input signal frequency ω_s, then the resultant output is

$$E_O = A \sin \omega_s t \sin \omega_c t \tag{1}$$

The trigonometric identity

$$\tfrac{1}{2} \sin (x + y) + \tfrac{1}{2} \sin (x - y) = \sin x \sin y \tag{2}$$

can be directly applied to (1), and E_O can be rewritten as the sum and difference frequency $(\omega_s + \omega_c)$ and $(\omega_s - \omega_c)$ or

$$E_O = \frac{A}{2} \sin (\omega_c + \omega_s)t + \frac{A}{2} \sin (\omega_c - \omega_s)t \tag{3}$$

If the input signal is 10 Hz and the carrier frequency to be modulated is 100 Hz, any electronics used to process the resultant signal must, in effect, amplify two sinusoids, one at 110 Hz and one at 90 Hz. The higher the input signal frequency, the more separated the two resultant modulated signals become.

When two signals are superposed on the input, each acts independently if the modulation process is linear. Nonlinear processes cause other sum and difference terms to appear that can be classed as cross modulation.

If a sinusoidal input signal is superposed on a dc signal, the dc level causes a static carrier level, and the ac signal produces a pair of signals above and below the carrier frequency. The resultant Lissajous butterfly pattern is shown in Figure 6.18a. Stated mathematically,

$$E_O = A_{dc} \sin \omega_c t + \tfrac{1}{2}A \sin (\omega_c + \omega_s)t + \tfrac{1}{2}A \sin (\omega_c - \omega_s)t \tag{4}$$

When two or more sinusoids are modulating the carrier signal simultaneously, the Lissajous pattern of Figure 6.17a still holds. The output signals are now just sums and differences for each independent signal.

---

[1] A restriction in bandwidth is necessary for this statement to have meaning.

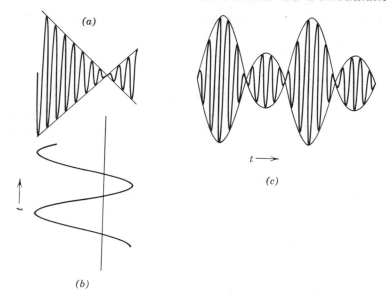

Figure 6.18 A Lissajous pattern for an ac and dc superposed modulation signal.

In carrier-suppressed modulation, the sum and difference frequencies should pass through the electronics if the input signal is to be properly processed. If the signals are placed through a transformer or an amplifier or a filter, the bandpass character of these elements directly influences the signal bandwidth. As an example if the carrier frequency is 1000 Hz, and the modulating signal is 200 Hz, and the filter passes signals only from 900 to 1100 Hz, then the 200-Hz signal will not be transmitted even though the bandwidth is 200 Hz. The 200-Hz signal generates two carrier signals, one at 800 Hz and one at 1200 Hz, and they are outside the bandpass of the filter. With the filter above, the signal bandwidth should be 400 Hz.

If the modulating element is a simple switch, the signal output is made up of square-wave segments. If the signal changes during the time the signal is connected, the modulated carrier signal will display this detail. This is shown in Figure 6.19 for a full-wave modulator. If the modulation is half-wave, the changing signal occurs only on part of the modulation output as shown in Figure 6.20.

The square-wave signals in Figure 6.20 are shown as ideal. Any amplifier processing this type of signal would have to have considerable bandwidth to reproduce this type of waveform faithfully. If the demodulation process is full-wave and perfect, the negative signals can be

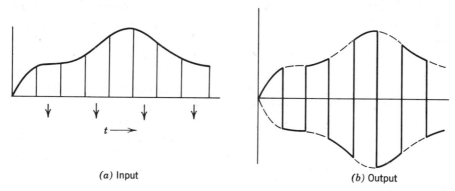

(a) Input   (b) Output

Figure 6.19   Full-wave modulated output signal waveforms.

"folded back" into the positive signals re-creating the original signal exactly.

In Figure 6.20 the negative segments are "flat topped." When these signals are folded back in a full-wave demodulator, the resultant is imperfect. The "fold back" gives an average value during alternate half-cycles.

A half-wave demodulator would contain a great deal more carrier-frequency content (ripple) than a full-wave demodulator as the output signal returns to zero each half-cycle. Such a signal would have to be more heavily filtered to reconstruct a facsimile of the signal being modulated. Half-wave demodulation of the signals of Figure 6.19 or 6.20 is thus about equivalent. Full-wave demodulation in Figure 6.19 is more accurate than that of Figure 6-20 at higher signal frequencies with a significant reduction in carrier ripple over half-wave demodulation.

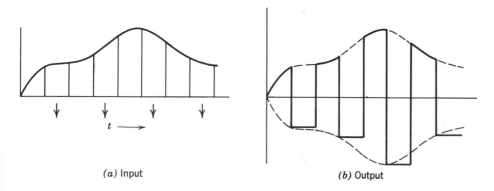

(a) Input   (b) Output

Figure 6.20   Half-wave modulated output signal.

When carrier signals are influenced by filters, transformers, amplifiers, or practical modulators, the perfect signals of Figures 6.19 and 6.20 are not realized. The electronics introduces phase shift for the higher harmonics making up the square waves. These signals are usually rounded or smoothed and delayed by the electronics. The timing in some modulation methods precludes an accurate fold back. This implies that an ideal demodulation process is difficult to achieve.

When ac signals at the carrier frequency are demodulated, the result is a dc signal providing the signal has a component in phase with the demodulator. If this signal was parasitically coupled into the signal stream, the resultant output signal would not correspond to an input signal and would be an error.

If the signal being modulated has frequency content near the modulation frequency, the demodulated output can contain this difference signal. If the modulation frequency is 1 kHz and the input signal has a 990-Hz component, a 10-Hz signal can appear in the demodulation. An error of this type is called an "aliasing error." To avoid such errors the input signal should be filtered to remove content above one-half the modulation frequency.

The frequencies making up a 1-kHz carrier signal modulated at 10 Hz are symmetrical about the modulation frequency. If an incoming signal of 990 Hz is modulated, as above, the symmetry does not hold. If the demodulation process on the 990-Hz signal were full-wave and perfectly symmetrical, no 10-Hz difference signal would appear. Aliasing error is thus a strong function of modem symmetry. Since symmetry cannot be guaranteed in general, the modulation of signal content above one-half the modulation frequency should be avoided.

6.9 NOISE AND CARRIER SYSTEMS

The frequency range of interest in any dc amplifier using modem techniques centers about the modulation frequency. The demodulator produces a dc output for center frequency content and ac output for signals located symmetrically about the center frequency. The demodulator treats all signals located about the center frequency including noise.

If the center or modulation frequency is 1 kHz, noise at 990 and 1010 Hz reduces to 10-Hz noise in the demodulated output. If the noise spectrum of the modulated signal includes a 20-Hz component, this signal would appear as 980 and 1020 Hz in the demodulated output. Since modulation content and its harmonics are usually filtered out, noise in this region is not of importance. Noise in the original signal is transformed to sidebands about the modulation frequency. Noise injected after the modulator is added to the transformed signal and only that noise centered about the modulation frequency need be considered. Noise developed by the modulator is usually related to the transformed signal and therefore is as important as the noise present in the input signal.

Noise transformation is one of the principal reasons for using modulators in the amplification of low-level signals. Noise developed by an input stage is usually higher at low frequencies (1/f noise, for example) than at the carrier frequency. If the input stage adds to the noise after modulation, only that content centered about the carrier frequency affects the S/N ratio. Thus it is desirable to use postmodulation gain rather than premodulation gain. In critical applications, operating points of input stages can be set to yield minimum noise at the desired carrier frequency.

6.10 CARRIER-SUPPRESSED MODULATION AND FEEDBACK IN DC AMPLIFIERS

Feedback can be applied to carrier systems in a variety of ways; for example, the amplifiers themselves can be feedback-stabilized or the feedback can include the modulation/demodulation process. It is possible to have a voltage null on the carrier signal or a null on the actual input signal. Transformers can be used both in the forward gain path and in the β loop.

Figure 6.21 shows feedback applied back to the input transformer primary. When the center tap of the input primary transformer is at the average signal potential of the input E_{IN}, no net ac signal is produced by the modulator. The output voltage attenuated by the divider R_1 and R_2 at balance equals E_{IN}, and this defines the gain as the ratio

$$\frac{E_O}{E_{IN}} = \frac{R_1 + R_2}{R_2} \tag{5}$$

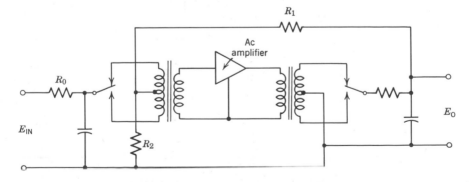

Figure 6.21 A modem with overall dc feedback.

This system is difficult to stabilize as two transformers are included within the feedback loop. The circuit results in a high input impedance with only the input filter outside of the feedback loop. The high input impedance results because very small signal currents flow to the transformer primary. If the center tap were exactly at the input signal level, no current would flow and the null impedance would be infinite.

The dynamic input impedance of the circuit in Figure 6.21 can, however, be very low. When the input signal changes, the center tap does not immediately change to its proper new level. During the interval of transition, input current flows. If the input signal is always changing, current is always flowing. The more rapid the change, the more current flows. This is exactly the characteristic of a capacitor. The input impedance of the amplifier in Figure 6.21 is that of a capacitor (excluding the input filter capacitor). The input impedance never drops below R_0, however.

To exclude the input transformer from the feedback path, the carrier signal can be fed back to the input transformer secondary as shown in Figure 6.22. The gain is defined by the attenuation factor $(R_1 + R_2)/R_1$ and by the turns ratio in the two transformers. The input impedance at null is limited by the magnetizing inductance of the input transformer. Note that the full input signal is placed across the primary transformer coils on each modulation cycle.

The principal advantage of the circuit in Figure 6.22 is the ohmic isolation between the input modulator and the output circuitry. Also the amplifier is easier to stabilize, since both transformers are removed from within the loop transmission. Ac feedback to the input transformer primary can be included without yielding on ohmic isolation. This circuit is shown in Figure 6.23. The gain of this system is determined by the ratio of resistors R_2/R_1 and by the turns ratio in the output transformer.

The preceding circuits all use transformers, which are desirable for full-wave processing and for impedance matching purposes. These circuits are only outlines and omit such important details as transformer

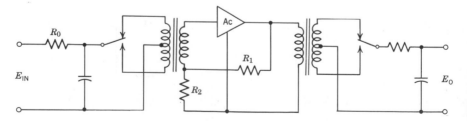

Figure 6.22 Ac feedback in a modem.

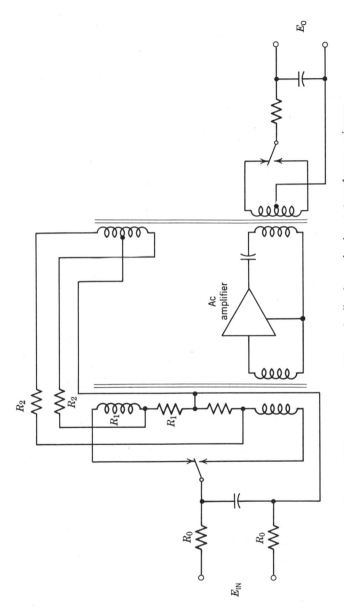

Figure 6.23 A modem with ac feedback to the input transformer primary.

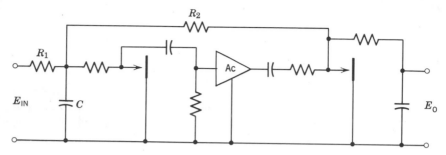

Figure 6.24 Operational feedback applied to a carrier-type dc amplifier.

shielding, optimum turns ratios, modulator drive detail, and stabilization elements.

Transformers are not essential in carrier-type dc amplifiers. One circuit example using operational feedback is shown in Figure 6.24. Capacitor C stores a charge that results from both input and output signals. When $E_0/E_{IN} = R_2/R_1$, the input summing point (voltage on C) is at null. (A finite null signal must be present to produce an output signal, namely, E_0/A.) Both the input and output switches are half-wave types and may actually be combined on the same reed structure. The input impedance to this type of amplifier is essentially R_1, which can be easily seen since the voltage across C is nearly zero. The potential drop across R_1 is simply E_{IN}, and by Ohm's law the input current-to-voltage ratio is just R_1.

The dynamic impedance of the circuit in Figure 6.24 is never lower than R_1. At high frequencies, the system can be slow to return the charge on C to zero. The input impedance thus tends to rise as the voltage on C follows the input signal. The input impedance now has two values: one with the modulator element shorted, and one with the element open. The modulated input impedance can be disturbing in some applications. Another view of this effect states that the modulator is "pumping" signal back into the source. Whatever the view, a modulated current does flow that is dependent on the "activity" of the input signal.

6.11   FEEDBACK AND NOISE REDUCTIONS WITH CARRIER-TYPE DC AMPLIFIERS

The feedback used in the preceding examples differs from the feedback described in Chapter 8 where modulation is not used. Those dc systems provide immunity from hum pickup or power-supply variation and define their own internal static operating points because changes in operating points are just signals being processed by the amplifier. In amplifiers with overall signal feedback, no feedback immunity exists against carrier pickup at the input.

In modulation schemes, the dc operating points are not necessarily controlled by the feedback as operating points and signal levels are not related. One technique used to define the operating points in an ac amplifier is to use decoupling and a feedback structure that sets the dc gain to zero while still leaving the ac gain at a maximum level. With the dc gain at zero, the operating points are fixed and cannot shift. One example of this technique is shown in Figure 6.25. The gain is the ratio of R_2 to $X_c + R_1$. At zero frequency, X_c goes to infinity and the gain goes to zero. At operating frequencies X_c can be considered a short circuit and the gain is simply R_2/R_1. When unwanted carrier is coupled into a modulation system, it is demodulated and the resultant direct current is added to the null point to modify the generation of carrier by the modulator. Thus the added carrier is modified (assuming it is in phase with signal carrier) just as direct current added to a dc feedback loop is modified. The point of injection determines the severity of any difficulty as the signal is reduced by the feedback factor and multiplied by the gain following the point of injection. Just as in a feedback dc amplifier, if the injection occurs at the input, the unwanted signal cannot be separated from desired signal and large errors can result.

When hum or noise is injected into a modulation scheme with feedback, the benefits of feedback are not always available. These signals are amplified in an open-ended sense. If their demodulation is outside the frequency band of interest, they can be ignored. If they are of sufficient amplitude to overload or distort a desired modulated signal, then the demodulated signal will be disturbed.

Noise added to a modulation scheme in the sideband region where signals are processed is demodulated. These signals have gain defined by the feedback process. Noise injected into the output stage of a carrier

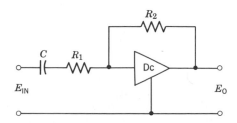

Figure 6.25   An ac amplifier with a dc gain of zero.

system will be reduced by the feedback factor if the frequency content is within the sidebands of the device. Noise outside of this range will not be reduced.

As stated earlier, carrier signal injected into a signal path will be modified by the existing feedback factors. This is *not* true if the carrier signal is out of phase with the phase of a normal signal at the same point. This deficiency can cause trouble if the signals are large and are of such a magnitude as to overload final amplifier stages. An example of this was brought out in the discussion on practical modulators when the problem of magnetic coupling at the contact closure was discussed. If any of the content is out of phase, it will not be considered an input signal; therefore a large ac component of out-of-phase (quadrature) signal can appear in the amplifier output. This signal level often determines the maximum practical gain that can be used before demodulation. Further amplification would simply force the last stages into constant overload.

In low-level modulation type dc amplifiers, the nature of the signals at the output of the ac amplifier requires explanation. These signals, stated simply, can "be a mess." Quadrature content and extraneous noise associated with the modulator (and demodulator too) make the resultant very difficult to interpret. If the signal levels are within normal operating bounds, the only criterion for judging performance involves the entire amplifier. Any attempt to judge performance within the modulation loop can be very misleading.

PROBLEMS

1. A sine wave is half-wave-modulated with a 40% duty cycle. Compare the average signal with a 50% duty cycle.

2. Using Figure 6.2c, tell what the negative peak signal is if the duty cycle is 30%?

3. In Figure 6.5 if the switch has a resistance of $0.1R$, illustrate the output waveform. What is the new value of E_P?

4. In Figure 6.8 assume a 400-Hz carrier and $R \ll R_1$. If $C_1 = 1$ μ F and $E_{IN} = 1$ V rms, what average voltage exists on C_1? Approximately what is the value of R_2 and R_1 to produce 1% ripple peak-to-peak?

5. In Figure 6.10 assume the modem is perfect half wave and the amplifier has gain 100. What is the dc gain?

6. In Figure 6.12 the transformer is 1:1 for the entire primary. If the secondary load is 10 kΩ, what current level is drawn from the the battery? Is it alternating or direct?

7. In Figure 6.13 the transformer is 1:1 for the entire primary. If the amplifier gain is 100 and the modem is perfect full wave, what is the dc gain?

8. Derive a formula for the frequency where the net dc gain is unity in Figure 6.25. If $R_2 = 100$ kΩ and $E_{IN} = 10$ V, what leakage resistance in C limits the dc offset to 10 mV?

7

Instrumentation Amplifiers

7.1 INTRODUCTION

The difficult problems encountered in instrumentation involve low-level signals. These signals arise from transducers such as thermocouples and strain gages connected to structures which are being measured for temperature or mechanical stress. Typical problems involve long input cables, the grounding of transducers, and dynamic measurements that involve the careful treatment of all types of noise sources.

A feedback amplifier by itself does not solve such instrumentation problems. The reasons are many and interrelated. Power-supply isolation, common-mode rejection, signal grounding, etc., all must be considered a part of the requirement. The problem thus extends beyond the instrument. Before the amplifier requirements can be stated, the system and signal specifications must be outlined.

7.2 THE FUNDAMENTAL PROBLEM IN INSTRUMENTATION

Consider a signal potential and its associated source impedances. The signal source can be electrically connected to the structure being tested. The signal potential difference is to be amplified and monitored at an instrumentation point. The problem is to amplify the signal potential difference while ignoring the potential difference between other parts of the system. Rack or framework potentials and the utility power neutral often influence the signal processes. The fundamental problem is to handle signals in the presence of all required earths, grounds, and structures.

Before a discussion concerning grounds and structures can be meaningful, the words that are used must be clearly defined. The words isolation, differential, and ground are not precise engineering words like ampere, collector, and voltage gain. They often have meanings which are related to prior experience, not to textbook definitions. The reader is cautioned to study the following definitions before reading further so that the "communications gap" will not be too great.

Definition. Earth. An ohmic connection to the soil by means of a buried water pipe, say.

Definition. Ground. Any conductive complex that connects to the zero reference signal conductor. Obviously included is an earth, an oscilloscope, the frame of an aircraft, or a piece of copper wire. Specifically, a ground point need not be an earth point.

Definition. Isolation. Not defined except in context. An isolated amplifier does *not* in itself convey specific information about the amplifier.

Definition. Shield (Electrostatic). A shield is a conductor geometrically positioned so that certain electrostatic flux lines are controlled or contained.

Definition. Zero Reference Potential. All voltage type signals of interest occur as potential differences. Since an absolute of zero potential does not exist, it is often convenient to select a reference point and refer to it as *the* point of zero reference potential. The selection of a zero point does not change the arrangement of potential differences. When discussing input signal processes, the zero point can be the connection to the input structure. When discussing signal recording, the zero point can be the low side of the signal connected in common to other signal conductors. Several zero reference potentials can be selected as long as they are each considered separately.

7.3 THE ELECTROSTATIC SHIELD

If an electronic device is fully enclosed in a metal box, external potential differences can cause current flow only in the surface of the box. Potential differences within the box are unaffected by the external electrostatic environment. In Figure 7.1, the potential E can vary and the potential differences within the enclosure are unaffected.

Mutual capacitances exist between all conductors in a shielded enclosure. The largest capacitances occur to the shield enclosure as it has

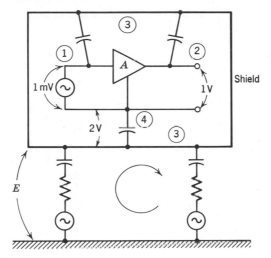

Figure 7.1   The electrostatic shield.

the largest area. Three specific capacitances shown in Figure 7.1 are
redrawn in Figure 7.2 for further exposition. Note that C_{13} and C_{23}
constitute a feedback path with C_{34} acting as an attenuator in the path.
Such coupling is undesirable and the usual practice is to attenuate the
feedback fully by placing a direct tie across C_{34}. In Figure 7.1 this
is equivalent to connecting the shield conductor to the zero reference
conductor of the amplifier.

7.4   SHIELD-TO-SIGNAL TIES

The need for a shield-to-signal connection becomes clearer if the zero-
potential conductor of the signal is grounded externally as in Figure
7.3. The potential difference between ground points E_{12} causes a voltage

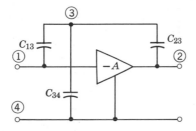

Figure 7.2   Capacitance coupling in a shield enclosure.

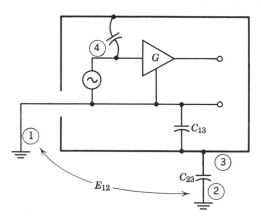

Figure 7.3 The problem of a floating shield.

division between capacitance C_{12} and C_{23}. Conductor ③ can thus couple a fraction of the potential difference E_{12} into point ④. The fraction goes to zero when capacitance C_{13} is replaced with a direct connection.

Current flow in signal conductors causes potential drops which are interpreted as signal. If the current is noise or a power-related phenomenon, the resulting signal is undesirable. To appreciate the problem further, a typical cable run of 100 ft of #20 wire has a resistance of 1 Ω in each line. A current flow of 1 μA causes 1 μV of unwanted pickup. Capacitances of 270 pF and voltages of 10 V at 60 Hz can cause 1 μA to flow. Typical amplifier designs involve capacitances to the shield much greater than 270 pF and potential differences (transformer voltages) much greater than 10 V.

Shield-current loops should not use a signal conductor to complete a

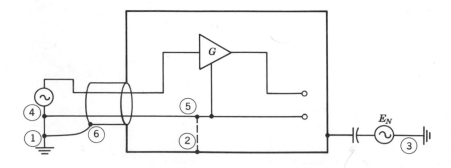

Figure 7.4 A proper input shield connection.

path. Thus shield-to-signal conductor ties should be made at the point at which the signal is grounded. The correct connection is shown in Figure 7.4. Current flowing from the voltage source E_N takes the path ③ ② ① ③. No current flows in conductor ④ ⑤. If the shield connection were made at ② ⑤, current from E_N would take the path ③ ② ⑤ ④ ① ③. This path includes conductor ④ ⑤ and unwanted pickup would result.

7.5   ELECTROSTATIC SHIELDING RULES

1. All conductors carrying signal potentials should be carried inside a shield conductor.

2. The shield conductor should be connected once to the zero-reference conductor of the signal.

3. Shield-to-signal connections should be made at the point at which the signal conductor grounds to its external environment.

7.6   SHIELDING DISCUSSION

The above rules are complete, however, a few additional comments will help to make them practical. It is not necessary to obey the rules if the penalty for violation has been carefully considered.

Low-level high-impedance circuits require the most careful attention to shielding. High-level low-impedance signals can go unshielded as long as coupling to other low-level signal points does not occur. In many cases the signals are shielded by their natural electrostatic environment and additional or separate shielding is superfluous; for example, a thermocouple bonded to a structure has its leads shielded by the structure.

It is unnecessary to shield an area if no contaminating influence will exist. Since transformer shields may prevent coupling to transformer coil potentials, they are required within the transformer. If coil voltages do not occur at other points, it may be unnecessary to shield these other points.

In a circuit with one zero-reference conductor, there is only one correct shield potential. A *single* connection of a shield conductor will ensure that this potential has been defined. Multiple shield connections are unnecessary and are usually undesirable. If a shield connection is made at two points, current will flow in the shield establishing a potential gradient. If one end of the shield is correctly defined, all other points will be incorrect and this will couple unwanted signal to the internal circuits.

If a signal is processed through several interfaces, the shield potential must be carried through as a separate conductor. Shield conductors in separate areas cannot be independently or indiscriminately connected to a convenient ground. When a shield conductor is connected at two points, it is only to provide continuity, not for the purpose of permitting current flow.

Shields and signal conductors are often in the form of two-wire shielded cable (twinax). Two shields over a single conductor, triax, can also serve the same purpose although preparing the shields for termination is more difficult. If coax is used, the outer conductor serves to shield the center conductor but this is not the separate type of shielding that is discussed above. Coaxial cable can be used in high-level signal transfer as signal contamination can be neglected. Coaxial cable for low-level signal lines is not recommended as parasitic current must flow in a signal-carrying conductor.

7.7 INSTRUMENTATION DIFFERENTIAL PROCESSES

The fundamental instrumentation problem can now be discussed in terms of the shielding rules. Consider a signal developed at one ground point and observed at a second ground point as in Figure 7.5. The signal has two zero-reference conductors. The electronics which processes the signal can be divided into two regions, each region shielded by a conductor connected to its signal at its ground point. The signal is properly handled if the contribution of E_{13} is a small part of the output signal.

The leakage impedance Z of Figure 7.5 permits current flow in path ① ② ③ ④ ① if there is a potential difference E_{13}. If R is 1000 Ω and E_{13} is 10 V and the unwanted pickup is to be less than 10 μV, then the current level in Z must be below 10 nA. Thus Z must be 1000 MΩ or greater. At 60 Hz this value is 2.7 pF—not much leakage capacitance. Note that Z is a generalized impedance that can permit current

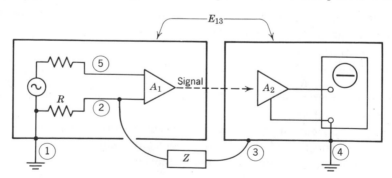

Figure 7.5 The two-shield problem.

to flow in the source impedance. No statement was made about the gain of A_1 or A_2.

The voltage difference $E_{21} - E_{51}$ is the signal to be amplified. At the output, point ④ is the zero-reference conductor. Potentials E_{24} and E_{54} both include the potential difference E_{13}. If $E_{21} - E_{51}$ is zero, then $E_{24} = E_{54} = E_{13}$. As viewed from point ④, the potential difference E_{13} is common to both input signal lines. Thus potential difference E_{13} is a common-mode signal. Stated another way, *the ground difference of potential is a common-mode potential for signals processed between two ground points*. If a common-mode signal E_{13} of 10 V results in an unwanted signal at the input of 10 μV, the common-mode signal is said to be rejected in the ratio 1 million to 1. Stated in decibels, the common-mode rejection (CMR) ratio is 120 dB.

7.8   CROSSING THE ELECTROSTATIC BOUNDARY

Signals can be transferred across an electrostatic boundary magnetically, electromagnetically, electrically, or mechanically. A magnetic transfer system might include a modulation process, a transformer, and a postdemodulator. This scheme is theoretically perfect as no holes in the electrostatic shield are necessary. The system is often used when high-voltage common-mode signals might be encountered.

Electromagnetic transfer could include optical or rf links. These methods are not generally found within the confines of one instrument. Obviously an rf link in the broadest sense is a differential process with essentially infinite CMR. For economic reasons other techniques are used to handle the common-mode or two-ground signal-processing problem.

Direct electronic connections are practical. As previously indicated, the impedance permitting leakage current between the two enclosures must be held to a very high value, typically 1000 MΩ. Such an impedance level is difficult but possible to attain. The advantage of this approach is low cost—note that signal transformers are not needed. The limitation most often encountered involves reduced accommodation to common-mode signal levels.

Mechanical means of coupling from an electrostatic enclosure is mentioned only for completeness. This type of electrostatic boundary crossing is not used and is not discussed in subsequent sections.

7.9   POWER ENTRANCES INTO ELECTROSTATIC REGIONS

The problem of power entrance includes the proper treatment of transformer shields. If the shielding is correct, transformer coil potentials will circulate reactive currents in shield conductors—not in signal-carrying conductors. The shield treatment in Section 2.8 shows that two shields are required as a minimum to control the current flow. A primary shield is connected to a power neutral point and a secondary shield is connected to the signal common or power-supply common of the device being powered. Each electrostatic enclosure is theoretically connected once to the zero-signal-reference conductor at the point at which the signal conductor makes its ground connection.

For a single-ended amplifier a single electrostatic enclosure is required. In practice metal shield segments surrounding the amplifier are connected to the nearest zero-reference conductor. If the conductor is carried a long distance to its connection, potential drops along the conductor caused by parasitic current flow can couple high-frequency noise into the instrument. The usual segmenting of shield conductors is shown in Figure 7.6.[1] The local metal framework is connected from ② to ③. The signal shield is connected through from ④ through ⑤ to ⑥. Note that the single-point shield connection to ground is still made at ④. Secondary reactive energy circulates in path ⑩ ⑪ ⑫ ⑩, while primary current circulates in path ⑦ ⑧ ⑨ ⑦.

If a potential difference exists between point ⑨ and ④, current can circulate in capacitance $C_{9,12}$ in the loop ⑨ ⑫ ② ④ ⑨. This current unfortunately flows in the input signal conductor ② ④. If shield ⑨ is carried back to point ④, the ground difference of potential will cause current flow in conductor ④ ⑨, not in ② ④.

Another source of potential can exist in loop ⑨ ⑫ ② ④ ⑨. It results from any magnetic flux intercepted by the loop from power lines or transformers. To eliminate such flux capture, loop area ⑨ ⑫ ② ④ ⑨ can be made minimal by running the connection from ⑨ to ④ very close to the input cable. Another possible solution uses the input shield conductor itself to connect point ⑨ to point ④. If the shield current is large, this practice is not recommended.

Figure 7.6 is a worst-case situation as an input signal point is grounded. If the output were grounded, the current flow would be the same but in an output signal conductor. Obviously this is far less critical as the potential differences that result are not subject to amplification. The practice of connecting the primary shield to the output ground is correct but unnecessary unless low-level signals are being conditioned.

The signal region around ④ in Figure 7.6 is shown open. If the source is guarded by the mounting framework, additional shielding is superfluous. The output is shown completely enclosed—for example, a galvanom-

---

[1] The iron-core designations have been omitted for clarity.

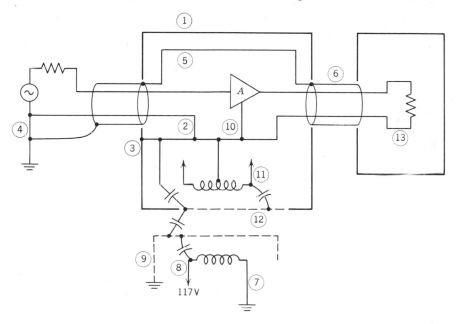

Figure 7.6 Practical electronic shielding of a single-ended instrumentation amplifier.

eter or meter. If point ⑬ could couple reactively to an outside ground, current would flow in the entire signal conductor path ④ ② ⑩ ⑬. If the output is grounded, then the input segment requires a complete shield enclosure. The shield must *not* connect at the signal source but at the output ground point per the shielding rules.

If the zero-reference conductor of the signal source is resistively connected to the zero-reference conductor of the amplifier, the shielding shown in Figure 7.7 is correct. Three transformer shields are preferred although the primary shield could be omitted. In this problem, shield ① is connected to the source ground and shield ② is connected to the zero-reference conductor of the amplifier. The additional primary shield ④ keeps primary circulating current from flowing in the input shield. If the shield scheme is not correct, small currents circulating in R can result in very large signal contamination.

Two enclosures ⑤ and ⑥ are shown connected to shields ① and ②, respectively. Shield ⑥ need not be complete except near critical input points. Shield ⑤ is often called a "guard shield" and should enclose all the instrument. The tightness of shield ② in the transformer is directly related to how much current can circulate in leakage capacitance C_{17}.

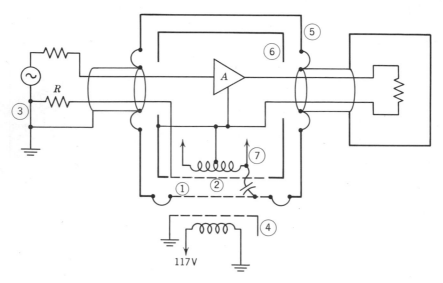

Figure 7.7 A three-shield transformer application.

If R is 1000 Ω, C_{17} should be below 1 pF. The circulating current from secondary coil potentials follows the loop ② ⑦ ① ③ ②, and 60-Hz current in R is input signal for the amplifier.

7.10 POWER ENTRANCES INTO THE DIFFERENTIAL AMPLIFIER

The double electrostatic process of Figure 7.5 requires separate power into each electrostatic domain. This power can enter both regions from outside or transfer between regions after being supplied into one section. Utility power or carrier power can be involved in various combinations, but the two electrostatic domains must still be protected. The single transformer external utility power solution is shown in Figure 7.8. Note the separately shielded secondaries. The shield connections are not shown in detail (see Figure 7.6) as the intent is to show only the shielding philosophy.

If power is carried between the two regions, the solution is shown in Figure 7.9. This power can be at a higher carrier frequency or at 60 Hz, depending on other specifications. Again, three shields are required to transfer power from the output domain to the input region.

In the following sections signal transfer in various differential amplifier configurations is discussed. The specifics of power shielding are

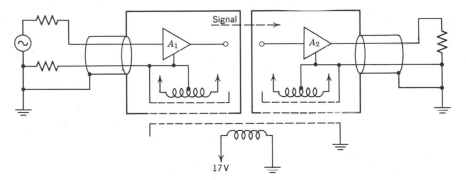

Figure 7.8 Power entrance to a double-shield system.

omitted in most cases for brevity. In each example a solution given in this section will be applicable.

7.11 DIFFERENTIAL INSTRUMENT AMPLIFIER DESIGNS

The two shield regions shown in Figure 7.5 indicate two separate regions of circuitry. In some designs the input leads are the only elements in the input shield region. In other approaches, only passive elements occur in the output region. The design approach that is selected is a function of cost, specification requirements, and the availability of suitable components.

When transformers are used to couple signals, high common-mode voltages up to 500 V can be accommodated. Leakage through the coupling transformer must be held to a few picofarads to achieve a meaningful CMR ratio. If direct coupling is used, common-mode signals must be processed by the electronics. If signal transformers are disallowed, high-voltage common mode can be accommodated by the use of properly placed attenuators. This procedure adds power-supply cost as separate power sources are needed for each block of gain.

CMR ratio varies as a function of design. A CMR ratio value is calculated by noting the error at the amplifier output. The ratio of error signal to common-mode signal times the gain is the CMR ratio value; for example, if the gain is 1000 and a 10-V common-mode signal causes 5-mV output error, the CMR ratio is given as $1000 \times 10\,\text{V}/0.005 = 2 \times 10^6$, or 126 dB. If G is the gain and E_{CM} is the common-mode signal and E_R is the output error signal, the CMR ratio is given by

$$\text{CMR} = 20 \log \frac{E_{CM}}{E_R} G$$

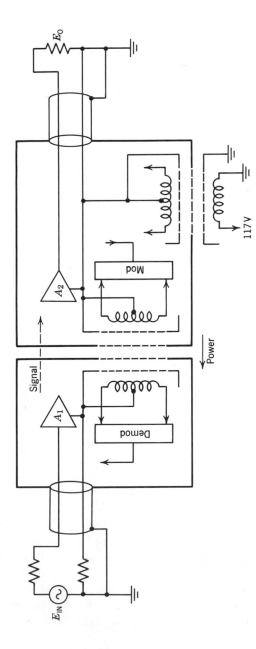

Figure 7.9 Internal power transfer.

In designs using transformers the CMR leakage path permits common-mode current flow in the input source unbalance. Here the error and gain reduce together and the CMR ratio stays constant with gain. In designs in which interstage CMR is used the CMR ratio is highest at highest gain and tends to fall off as gain is reduced. This is not harmful in most instances as the signal level is usually proportionately greater if lower gains are required.

In designs using feedback technique for CMR the CMR ratio is often directly proportional to gain. If the gain is reduced and the common-mode error stays constant with level at the output, the CMR ratio referred to the input becomes a smaller number.

If input attenuators are used, the CMR ratio is lessened by this attenuating factor. If the CMR ratio is 60 dB at gain 10 and a 10:1 attenuator is used to obtain a gain of 1.0, the CMR ratio drops to 40 dB referred to this attenuated input signal.

7.12   INPUT MODULATOR

The modulation techniques described in Chapter 6 can be used to transfer signals between two regions as shown in Figure 7.5. The boundary is crossed with carrier voltage proportional in amplitude to the signal level. Figure 7.10 illustrates a system without feedback although the feedback arrangements shown in Chapter 5 are possible if care is taken to shield the necessary feedback paths. (Note that an input filter is required to eliminate -aliasing errors.)

The quality of operation is greatly dependent on the chopper that is used. Any noise or erratic operation here will appear as noise in the output signal. Mechanical chopper frequencies can be as high as 800 Hz, permitting bandwidths which approach 200 Hz. The use of a 60-Hz modulation frequency is often avoided because any carrier pickup is demodulated and appears as offset error. One common practice is to use a 30-Hz carrier so that the 60-Hz phenomenon is second harmonic and has negligible influence on performance.

FET and photoresistive choppers have also found wide application. Since the modulator drive signals are easily coupled into the signal circuits, designers must carefully treat this source of trouble.

An input transformer permits optimum impedance matching to input amplifier input stages. For narrow-band applications this fact permits a low-noise design. Dynamic input impedance is likely to be low but this characteristic is usually not an application problem. In general the very lowest drift amplifiers are designed using mechanical choppers and a suitable stepup transformer.

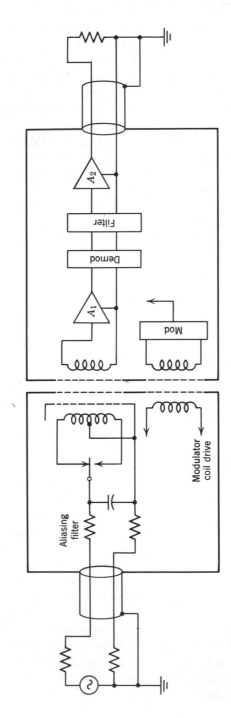

Figure 7.10 A mechanical modulator differential amplifier.

7.13   POSTMODEM WITH FILTER

If the electronics in the two regions of Figure 7.5 are arranged so that all the gain is in the input region, the design is that shown in Figure 7.11. The input amplifier can be single-ended as shown or treated in a balanced manner. The signal shield boundary occurs after the gain block with advantages in common-mode rejection, as discussed in Section 8.22.

The modulation drive signals can arise from either signal shield region or from an external power region. The carrier frequencies used are usually in excess of 50 kHz, and carrier transformer shielding is a critical problem. Carrier leakage can circulate carrier current through the signal source impedance causing regeneration and offset as discussed in Section 7.20.

If postelectronics are eliminated, power need enter only the input shield region. The low-frequency output impedance of this type of amplifier can be held to 10 Ω with efficient demodulator switches, and filter inductors are in the low mH range.

The circuit of Figure 7.11 uses a single-ended dc amplifier for gain. For a moderate source unbalance R_1, the shield capacitance C_{12} will not affect performance; for example, if $R_1 = 1$ kΩ and $C_{12} = 1000$ pF, the 3-dB frequency is 160 kHz. Such an unbalance (impedance to the input shield from either signal line) is troublesome in a few applications. For long input lines, the capacitance is relatively unimportant as it is only a part of a much larger shield-produced capacitance.

The single-ended amplifier approach of Figure 7.11 can be chopper stabilized. This type of stabilization is discussed in Section 7.17. Balanced solutions usually rely on matched differential input stages without chopper stabilization. Temperature effects can be treated in a variety of ways. (See the discussion in Section 8.21.)

Overall feedback is possible in this scheme, but to maintain the same shielding arrangement, the feedback path must involve a second magnetic-flux path. Since the quality of the β loop dominates the performance, the resulting feedback circuit is no better than the quality of the added magnetic coupling path. It is thus adequate to rely on the quality of one coupling in the forward gain path as no advantage is gained through the addition of a second feedback coupling path.

7.14   MODEMS WITH POSTAMPLIFICATION

Figure 7.12 shows an internal modem and postelectronics. This design has the advantage of a low output impedance and possible high output

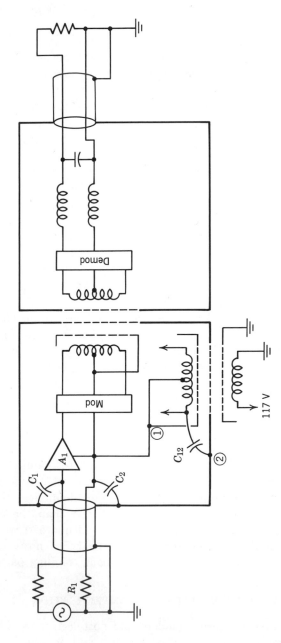

Figure 7.11 A postmodem differential amplifier.

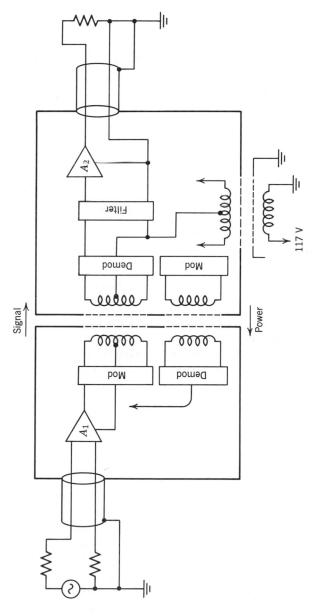

Figure 7.12 An intermediate modem differential amplifier.

current. Amplifier A_2 can actually be several parallel amplifiers which may provide ohmically separated outputs or a filtered output as application may dictate.

For optimum common-mode performance amplifier A_1 usually contains all the voltage gain. The modem can, however, be a current- or voltage-transforming device. Current transfer has the advantage of greatly reducing carrier coil voltages which can circulate signal carrier in the signal source impedance.

7.15   INPUT DIRECT COUPLING

Figure 7.13 shows a frequently used differential amplifier approach. Conceptually the idea of taking differences at the input terminals is appealing. (This approach parallels the approach used by oscilloscope manufacturers.) The input shield region contains only the input lines. The shield is carried into the amplifier as far as practical to eliminate any possible leakage capacitances to output ground.

The input impedances of the amplifier, Z_1 and Z_2, parallel any shield leakage capacitances C_1 and C_2. These impedances determine the CMR ratio directly. Input impedances from zero frequency to 60 Hz should be 1000 MΩ or greater to maintain a 120-dB CMR from a 1000-Ω source unbalance. The high input impedance must be present for CMR purposes and in no way implies that the amplifier can be used with high source impedances. Performance from high-resistance sources is limited by source currents required by the amplifier over its operating temperature range. These currents flowing in the input resistance appear as an offset or as drift that varies with temperature.

Input bias or "pumpout" current, even if only 1 nA, must have a

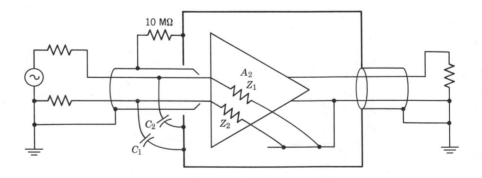

Figure 7.13 Input difference differential amplifier.

return path for loop completion. If current flows in both input lines, the only return path is from source ground to output ground. If the source and output are ohmically connected, no problem exists. If the source is floating at zero frequency, the 1 nA will block the input stages unless an alternate return path is provided. A common solution is to place a resistor between the two signal shields. Common-mode signals cause a few microamperes of shield current to flow and a bit more current is not harmful. If this current does not flow in the source unbalance, it is not sensed and amplified. A typical resistance value range is 1 to 10 MΩ.

The term *common-mode impedance* is often used. It is a measure of that impedance which permits common-mode current to flow in the source impedance. A simple ohmic measurement between grounds may yield a resistance related to a return-path requirement. This measure is not related to common-mode impedance which is more accurately determined by a CMR ratio test. Amplifier signal performance as a function of superposed CM voltage is critical in this type of design. If the CM signals are high-frequency in nature, slewing problems can severely distort normal signals. Note that at a gain of 1000, normal signals can be 10 mV, whereas CM levels may be 10 V. Also note that sustained CM signals can affect the heating of input stages with resultant long-term dc shifts.

7.16 INTERSTAGE DIRECT COUPLING

The differential stage with high-input impedance for CMR can be in a second amplifier; the configuration is shown in Figure 7.14. Amplifier A_1 can be single-ended or balanced as illustrated in the figure. This technique permits a low-drift input stage design that is not complicated by the added problem of rejecting CM signals. The CM impedance

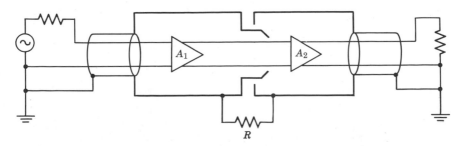

Figure 7.14 Interstage difference differential amplifier.

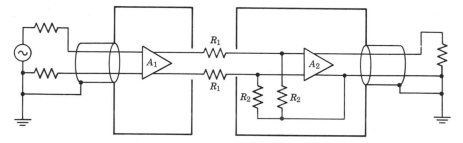

Figure 7.15 An interstage common-mode attenuator.

should again be 1000 MΩ, but the input amplifier may have a much lower figure. The return path for the input stage of the second amplifier must still be provided by a shield-to-shield resistor. This resistor is R in Figure 7.14.

A very useful circuit that attenuates the common-mode signal at the intertie between amplifiers A_1 and A_2 is shown in Figure 7.15. The attenuator reduces normal signals as well as common-mode signals. Amplifier A_2 must make up the gain loss. Since the two attenuators R_1 and R_2 are equal and of moderate value (100 kΩ), the CM impedance problem of Figures 7.13 and 7.14 does not exist at the intertie. Also note that amplifier A_1 does not handle the full input-to-output common-mode signals. If the attenuator is 30:1, a ground difference of potential of 300 V presents only 30 V of common-mode signal to A_2. If A_2 has a gain of 30, the signal loss is regained.

The CMR ratio of A_2 from a balanced source can easily be made 30,000:1. With a gain of 30, this permits the CMR ratio to appear as 1000:1 at the output. For a gain of 1000 the CMR ratio is 10^6. This figure assumes no CMR contributions from the input of amplifier A_1. If a contribution occurs from both sources, the two errors must be measured at the output and referred back to the input as a function of gain setting.

7.17 SIGNAL COMMON MODE

Signals can arise as a difference between two voltages located in the same shield region. The potential difference between the zero-reference conductor can be very large compared with the signal difference. A typical situation is a four-arm bridge excited with ac or dc voltages as shown in Figure 7.16. Voltage E_1 could be a zero-impedance source and the balance could be made by just two elements, Z_3 and Z_4. Four equal

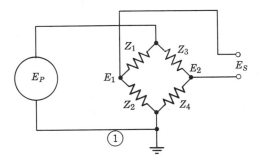

Figure 7.16 A typical bridge signal.

resistors constitute a typical strain-gage bridge. With $E_P = 10$ V, the voltage at E_1 and E_2 is 5 V. The difference signal $E_1 - E_2 = E_s$ can be full scale 10 mV. As viewed from ① the CM signal between E_1 and E_2 is 5 V; that is, $(E_1 + E_2)/2$. This is not the same as the CM signal as viewed from a second ground reference point. The latter signal when present must be rejected to obtain useful information.

The instrumentation problems encountered in design are discussed below. Figure 7.17 shows a typical resistance bridge and a signal amplifier. Capacitance C_1 and C_2 of the input cable shunts the signal sources E_1 and E_2. If the bridge is excited by an ac voltage, the capacitance must be a part of the reactive bridge balance. If the bridge is excited by a dc voltage, high-frequency performance of the bridge is valid only if a reactive balance exists. For a 1000-Ω bridge excited by a dc voltage a cable capacitance of 10,000 pF per line attenuates the signal 3 dB at 16 kHz. If full-scale signals are 10 mV for 10 V excitation, the reactive

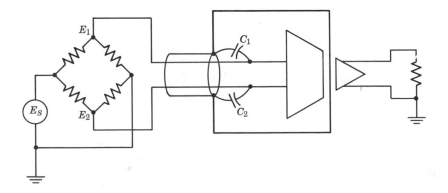

Figure 7.17 An amplifier and a bridge signal.

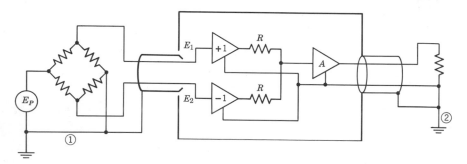

Figure 7.18 The double common-mode problem.

load should be matched to 0.1% or 10 pF to maintain signal common-mode balance at 16 kHz. At 3.2 kHz the capacitance balance needs to be held to approximately 100 pF. If the bridge is excited by an ac voltage, the dynamic response is further complicated by the presence of signal sidebands. Unequal treatment of the sidebands disturbs the signal.

Amplifiers that reject CM signals actively by having a high input impedance often have a limited range to the signal levels E_1 and E_2. The total CM signal as viewed from output ground is E_1 and E_2 plus the ground difference of potential. If the maximum CM signal is 10 V, a 5-V value for E_1 and E_2 limits the peak ground difference of potential to 5 V. This situation is shown in Figure 7.18 in which two amplifiers are shown taking the signal difference. (This circuit is impractical and is shown only for conceptual reasons.) If the potential difference ① ② is E_{CM}, the signals from ② at the amplifier inputs are $E_{CM} + E_1$ and $E_{CM} + E_2$. In taking the differences, the output should be proportional to $E_1 - E_2$ so that both CM effects, E_{CM} and $(E_1 + E_2)/2$, are rejected.

The measuring amplifier can be of a type not sensitive to the average value $(E_1 + E_2)/2$. One example is the postmodulator type shown in Figure 7.11. The average potential $(E_1 + E_2)/2$ stores energy on the capacitors C_1 and C_2 but is not used as a functioning signal in amplifier A_1. For this reason it can be any voltage within the voltage breakdown limits of the insulation used.

The capacitances C_1 and C_2 are likely to be unequal in a design as shown in Figure 7.11. If the input amplifier A_1 has symmetric high-impedance-input terminals, then capacitances C_1 and C_2 do not include isolation and power-transformer capacitances. This consideration is important if high-frequency dynamic content is being observed from the signal source. It is also important if the source is excited by an ac

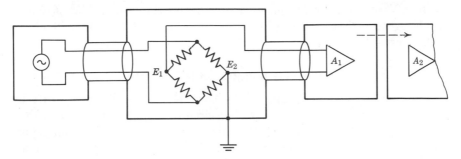

Figure 7.19 A floating source and amplifier.

voltage. The circuit of Figure 7.11 would be undesirable because of the reactive unbalance present.

The input shield should not usually be tied to one side of the signal unless the excitation source is floating. Such a circuit is shown in Figure 7.19. Note that the average value $(E_1 + E_2)/2$ is zero when the shield is connected to one side of the signal. This would seem like a desirable method of reducing common-mode level at the input stage. If the source is not properly isolated, unwanted current will flow in one arm of the source and add to the normal signal as shown in Figure 7.20. Potential differences to ground along the cable run or between grounds will circulate current in R; for example, current will flow in loop ① ② ③ ① for a potential difference E_{13} and current will circulate in loop ① ② ④ ⑤ ⑥ ① through C_{45} and R from a potential difference E_{16}. If the grounds and cable runs are very short and if the design has been thought out so that the potential differences are zero or are unimportant, then the arrangement of Figure 7.20 can be used.

The CMR ratio for signals on a resistance bridge must be as high as the CMR ratio for ground differences of potential. If a bridge signal

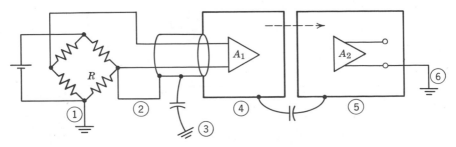

Figure 7.20 An improper shield configuration.

is 10 mV and the bridge excitation is 20 V, the bridge CM signal is 10 V. A 10^6 CMR ratio reduces the 10-V CM signal to a 10 μV or a 0.1% error.

7.18   "CHOPPER STABILIZATION"

Operational feedback gain blocks can be stabilized for dc drift by the addition of a modem. The added circuitry is similar to that shown in Figure 6.11. By paralleling two feedback loops, the modem loop defines the very low frequency performance. The high-frequency performance not possible from the modem alone is provided by the alternate parallel signal path.

At an operational feedback summing point, the only proper currents arise from an input or output signal. If a modulator senses any dc offset appearing at this point, an amplified version can be used to feed-back a correcting dc signal into the dc loop. For a zero input summing-point current, a zero potential at the summing point is possible only if the output voltage is zero. Note the first stage of amplification can be a source of summing-point current. This source can be eliminated by adding a coupling capacitor C_1 to the input. The capacitor makes the wideband dc amplifier an ac amplifier and places the entire burden of dc gain on the modem.

Figure 7.21 shows a typical parallel feedback structure. In some de-signs A_1 is simply the input differential stage. In other designs A_1 and A_3 constitute the major part of the open-loop gain, and A_2 is a final gain of 100 for both the low-frequency and high-frequency gain loops.

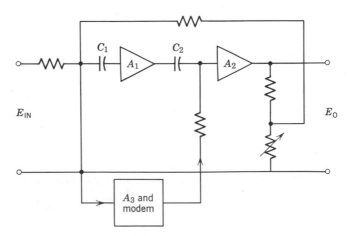

Figure 7.21   A parallel feedback structure.

At an intermediate frequency, signals are processed by both loops. Consideration is required to keep the gain constant in this transitional region. Gain errors in the transition area are termed "crossover" errors.

7.19   INTERMODULATION

The amplifier combination A_1, A_2 processes signals which include the modulation frequency used in the modem. These signals must have a finite summing-point error level. The modem amplifier also "sees" these error signals as if they were generated by the modulator. Their subsequent amplification and demodulation by A_3 produces an output signal at the modulation frequency and its odd harmonics. If these signals are exactly synchronous with the modulator, the net result is a dc offset. If the signals are near the synchronous frequency, a zero-beat phenomenon results. Such an output error is termed an *intermodulation error*. If output signals from amplifier A_2 are coupled inadvertently to the input amplifier A_3, very high intermodulation errors can result. This is usually a critical point in designs of this type.

Zero-beating intermodulation errors are highest at the modulation frequency and fall off at higher harmonics because only a fraction of one cycle can cause an intermodulation error. If the coupling is reactive, the pickup signal is higher at high frequencies which would normally offset the "falling-off" process. It is usual practice however to band-limit the modem amplifier for other reasons, and this action in turn again reduces the higher frequency intermodulation affects. The band limiting should occur at the input stages to eliminate any overloading possibilities from high-frequency summing-point errors. If these stages are ever overloaded, the error reflects on the stage capability to perform its assigned task at the modulation frequency and very strange low-frequency behavior can result.

The zero-beat intermodulation error reduces as the difference frequency increases. The error curve is directly related to the "crossover" frequency. If "crossover" occurs at a high frequency, intermodulation errors can cover a wider frequency spectrum. For this reason it is desirable to keep the "crossover" frequency low to reduce the region of frequency error.

7.20   CARRIER REGENERATION IN POSTMODEMS

The modulation products in a postmodem amplifier appear on the coils of the coupling transformers. These transformers are often toroids and shielding is a particularly difficult task. Adequate shielding is made

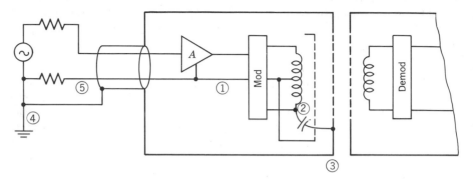

Figure 7.22 Reactive current circulation.

doubly difficult because the frequencies involved are above the band of interest and often include the harmonics associated with square-wave modulation.

Leakage currents flow in the input shield and in the source unbalance as shown in Figure 7.22. The path includes the loop ① ② ③ ④ ⑤ ① with voltage E_{12} and capacitance C_{23}. Signal sensed in the resistor R_{45} is amplified by A and can be remodulated. The result is a distorted output signal or even instability.

The amplifier gain can be reduced at higher frequencies to eliminate this effect. It is common practice, however, to squeeze as much bandwidth as possible from A, and the difficulty appears only when R_{45} is made large and at high gains.

If the phase shift through A is such that a 90° phase relation exists between the sensed signal and the modulator, the difficulty can be seemingly canceled out. Under these conditions a long input cable will modify the phase relationships and an unstable response can result.

If the amplifier has a balanced input, that is, no ohmic path from ① to ⑤ exists, the problem is somewhat modified. The voltage E_{12} acts as a local common-mode signal for the amplifier and the high-frequency CMR defines the regenerative process.

A stepdown transformer can be used to couple signals across the shield boundary. The added coils themselves can be used as shields. If coil voltages are low at the coil-shield interface, the circulating currents are proportionately lower. Center-tapped coils and/or reactive bridging (adding capacitors from coil to shield) can also be used to balance the reactive current flow. Current transformer coupling has the obvious advantage of keeping coil voltages very low.

Carrier signal circulating in the source unbalance can be sensed by an oscilloscope monitoring the signal source. This signal level should not vary out of bounds as a function of line unbalance and as a function of amplifier gain. The signal limits are often not specified. Practical limits are defined by their effect on the source or on other monitoring devices.

PROBLEMS

1. A gain 1000 amplifier has an output CM signal of 10 mV p-p for a CM drive of 60-Hz 20-V rms. What is the CMR ratio?
2. If the CMR is the result of leakage capacitance, what is the CMR ratio at twice the frequency in 1?
3. If an attenuator of 100:1 is used at the input, what is the CMR ratio referred to the new input? If the attenuator is at the output, what is the CMR ratio?
4. A 10-V peak-modulated signal at 500 kHz has a rise time of 0.3 μsec. What current flows in 10 pF of shield capacitance? Note that $I - C(dE/dt)$.
5. Design an input filter that is down 40 dB at 400 Hz to eliminate possible aliasing errors. Select an R to accommodate 1 μV of noise in 10-Hz bandwidth.

8

Specifications and
Evaluation

8.1 INTRODUCTION

The detailed qualities of an amplifier are apparent to the designer but often go unnoticed by the average user. It is possible to have an amplifier pass a full set of tests only to be useless in application. During much of the preceding discussion many typical design idiosyncrasies are pointed out. A person well aware of such possible shortcomings is better qualified to evaluate a design. Many specifications are implied and do not need evaluation; for example, it is unnecessary to look for 60-Hz intermodulation unless a 60-Hz modem is present. On the other hand, high-frequency CMR should be evaluated even though the specification is qualified at 60 Hz.

Specifications can be a "coverup" as well as a description of performance. The evaluation procedures, if not correct, can throw bad light on a good design. The best solution is to evaluate and measure with the best tools available, which includes a knowledge of what to expect and how to avoid pitfalls in procedure.

8.2 NULL TESTING

A powerful evaluation tool involves a technique known as null testing. Here the expected response is subtracted from the actual response. The difference signal is a residue of errors generated by the instrument or device. With care these errors can be measured and ascribed to individual specification parameters.

Negative-gain devices are amenable to null testing through the use of summing resistors. Positive-gain devices usually require the use of a floating meter or a differential oscilloscope. Techniques vary depending on whether the gain is greater or less than unity. Since much of the specification testing to be discussed can employ null techniques, it is important to discuss these circuits and their limitations in detail.

8.3   THE BASIC NULL CIRCUIT IDEA

If a signal is attenuated by a factor equal to amplifier gain, the amplifier output will be equal in value to the unattenuated signal. If these two signals are of opposite polarity and are summed, a null signal results. The circuit is shown in Figure 8.1. If $(R_1 + R_2)/R_2 = A$, then $E_0 = -E_{IN}$. The attenuator $R_1 + R_2$ provides a method of measuring $-A$.

The circuit of Figure 8.1 is conceptually simple. The circuit values must be selected carefully or serious errors result. The problem areas are:

1. Attenuator R_1, R_2 should be low enough in impedance so that parasitic capacitances or amplifier input impedances do not affect the attenuation factor.

2. Resistor R_2 should be a four-terminal type as discussed in Section 1.10. For gain 1000 R_2 may be 10 Ω. If an accuracy to 0.02% is desired, a 2-mΩ error resistance is required.

3. Amplifier output current flows in path ① ② ③ ④ ⑤ ⑥ ① and in

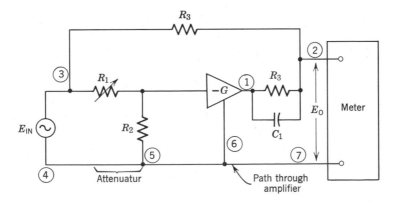

Figure 8.1   A null measuring circuit for negative-gain devices.

particular in conductor ⑤ ⑥. If this conductor is 2 mΩ, then $2R_3$ must obviously be greater than R_1 or the current in ⑤ ⑥ will produce a signal equal to the error limit for R_2. In general R_3 should be 10 times greater than R_1, and conductor ⑤ ⑥ should be kept as low in resistance as practical.

4. Only one connection between ④ and ⑦ is permitted. If the signal source and meter are both grounded, the null device will detect signals developed from ground-loop currents. It is advisable to isolate the input source from ground and use the amplifier connection ④ ⑦ as the generator ground. At high frequencies the reactive connection to ground through the power transformer does reinstate the ground loop. Fortunately many other factors disturb the null measure at high frequencies and this effect can be neglected.

5. Resistors must be free from voltage coefficients and matched for temperature coefficient.

The generator source impedance is not relevant in this circuit. If the source can supply sufficient voltage, the circuit will function. At null the signal level at ③ is equal to the signal level at ①. The signal generator E_1 can have a variety of waveforms. Sinusoids, square waves, ramps, and a dc voltage are all possible signals. Because of the nulling process, these signals need *not* be precise. If the sine-wave source has 1% distortion, the amplifier will follow the 1% error and still produce a null. Even if the amplifier has a 0.1% error response to the 1% distortion, the resulting error would be 0.001%, far below the limits of observation.

Sine-wave testing using the null circuit of Figure 8-1 assumes that the amplifier under test has zero phase shift. This is a valid assumption from zero frequency up to a few percent of the —3-dB point in most dc amplifiers and in a narrow midband range in an ac amplifier. To overcome this limitation, a trimming capacitor can be added between points ① and ②. The capacitor adds phase lead to the summing network equal to the phase lag of the amplifier. If the capacitor is placed between ② and ③, it can accommodate for phase lead below midband in an ac amplifier.

If the phase correction must accommodate more than a few degrees of amplifier phase lag, the calculated gain measure will be low. The gain error is proportional to $\cos \phi$, where ϕ is the corrected phase angle. The observed gain G_0 is

$$G_0 = G \cos \phi \qquad (1)$$

where $\Phi = \tan^{-1} \omega R_3 C_1$ and G is the correct gain value.

If $R_3 = 100$ kΩ, the summing-point source impedance is 50 kΩ. A metering point load of 1-MΩ loads the summing point ② by 5%. A 5% error on a 1% error measurement is usually acceptable; for example, if the linearity is actually 0.1%, the 1-MΩ loaded attenuator would indicate a linearity of 0.095%, well within acceptable tolerance. This summing point can also be affected by reactive loads such as a shielded cable. At 10 kHz a capacitance of 300 pF attenuates the signal by 30%. Often a capacitor is added to filter out spurious noise, but if the capacitance should attenuate an error observation, the result will be incorrect.

The chief advantages to null measuring techniques are the following:

1. Signal sources do not need to be of high quality or well calibrated.
2. Signal sources are always used at large signal levels.
3. Measurements are dependent on resistor accuracies, not on other electronics.
4. The full dynamic range of the instrument is observed at one time.
5. Only error signals are observed as the fundamental signals are nulled out.

If differential amplifiers are tested, they can be checked as negative-gain devices by grounding their positive input terminal. Since instrumentation devices have no through-ground connection, a jumper from ⑤ to ⑦ is necessary before the summing processes in Figure 8-1 can take place. The input-signal generator should still take path ⑤ ⑦ to ground rather than an unknown power ground return.

8.4   NEGATIVE GAINS BELOW UNITY

Low gains can be measured by interchanging the attenuator and the summing connection as shown in Figure 8.2. Because the gains are low, many of the critical factors discussed above are unimportant. The big

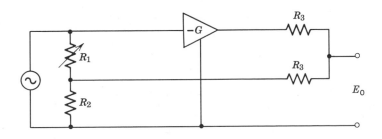

Figure 8.2   A low-gain nulling circuit.

problem is adequate signal level at the input. If the gain is 0.1, a 100-V input signal is required for 10 V of output signal. The dissipation and voltage coefficient of the input attenuator become important. If high-value resistors are used, the loading by the amplifier input must be considered.

8.5 POSITIVE-GAIN NULL TECHNIQUES

The summing circuit of Figure 8.1 is not applicable in a positive-gain amplifier null test unless a negative-gain amplifier has been intentionally inserted to invert the gain. Positive-gain null observations require direct subtraction between input and output points as shown in Figure 8.3. The points ① and ② can be connected to a floating meter or to a differential oscilloscope. At null the attenuator R_1, R_2 equals the gain value A. For sinusoids a small amount of phase lag in A can be accommodated by the added components R_3, C_3. Again the gain error will be proportional to $\cos \phi$, where

$$\phi = \tan^{-1} \omega R_3 C_3 \tag{2}$$

It is a temptation to use a floating single-ended meter for observation or an oscilloscope connected directly to points ① and ② with the low side connected to ①. This practice is not ruled out although a serious dynamic loading problem for the amplifier usually results. The observed error signal usually becomes very noisy, but if the error to be measured can be determined, the method must be considered adequate. It is not a recommended procedure because of the possible difficulties that can arise. One prerequisite is that the output impedance of the amplifier be below 1 Ω to keep the noise within bounds.

Amplifiers with positive gain below unity require a null circuit that interchanges the attenuator and observation point as shown in Figure

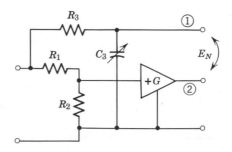

Figure 8.3 Positive-gain null measurement.

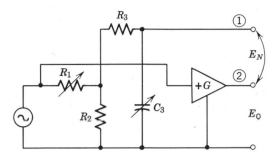

Figure 8.4   A null circuit—positive gains less than unity.

8.4. The high-voltage requirements for the input attenuator must be considered just as in Section 8.4. Elements R_3 and C_3 can accommodate for phase lag on the amplifier under test. The null points should be observed with a floating meter or a differential oscilloscope.

8.6   GAIN AND LINEARITY

Gain is defined roughly as the ratio of output signal to input signal. This definition becomes inadequate if noise, offset, and nonlinearity are included. A definition involving only differences does not include the measure of nonlinearity and noise although it eliminates any static or dc offset problems. A definition using the ratio of differences is an improvement, but this measure does not specify the degree of change or how a dynamic-gain measure might be considered. In the ideal case a gain measure obtained by differences is independent of the signal levels used but in practice other factors must be included in the measure. An all-inclusive definition is not possible. In the following sections the practical problems of measuring gain and nonlinearity are discussed to emphasize this point.

8.7   LISSAJOUS PATTERNS FOR NULL OBSERVATION

The ideal voltage gain is a simple multiplier on an input-signal voltage; that is, if $E_0 = GE_{\mathrm{IN}}$, then G is the voltage gain. Consider the null test circuit of Figure 8.1 with low-frequency sinusoids on the input-signal terminals ① ②. This same signal is applied to the x axis of an oscilloscope with the null output ③ on the y axis. The Lissajous figure that results can be used to detect a null measure of gain or linearity. The circuit is shown in Figure 8.5.

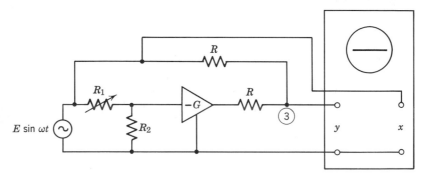

Figure 8.5 A Lissajous null-pattern circuit.

For an *ideal* amplifier with gain $-G$, the attenuator ratio is set to $(R_1 + R_2)/R_2 = G$, and at this setting, no y-axis signal is present. If the attenuator is set high by 1%, a 0.5% sinusoid will appear on the y axis, 180° out of phase. (A loss factor of 2 results because the summing resistors act as a divider on the difference output signal of the amplifier.) Figure 8.6 shows the null patterns for $+1$, 0, and -1% attenuator settings. Note that both x- and y-axis oscilloscope amplifiers must be direct coupled. The "zero tilt" setting of the attenuator thus measures the gain of the amplifier. Vertical displacement is simply dc offset of the amplifier or of the y-axis oscilloscope amplifier and is of no importance.

If the gain of the amplifier is now made 1% low for input signals above $+5$ V, the Lissajous pattern will display this difference as shown in Figure 8.7. The maximum error at the amplifier output is 1% of 5 V, or 0.05 V. After the 2:1 loss, it appears as a 0.025-V rise on the y-axis pattern at the right.

The attenuator setting R_1 can tilt the pattern in Figure 8.7 so that

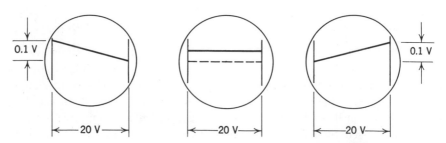

Figure 8.6 Lissajous patterns for $+1$, 0, and -1% errors in attenuator setting.

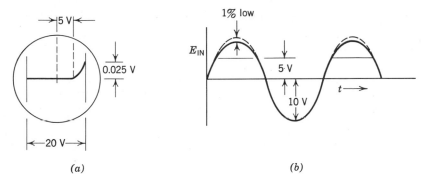

(a)   (b)

Figure 8.7 A 1% flattening of signal and a corresponding Lissajous pattern.

the peak y value is reduced. This pattern change does not move the center point at $x = 0$ but rotates the left side positive by about 0.013 V. The resulting pattern is shown in Figure 8.8. The largest peak error from a level Lissajous pattern is now 0.013 V. It would be correct to say that the peak error at the null does not exceed ±0.013 V and the peak amplifier error ahead of the summing divider does not exceed ±0.026 V. This implies that the gain error measure is ±0.26%, not 1% as in Figure 8.7.

The philosophical question posed is, which gain measure is correct, that of Figure 8.7 or 8.8? The latter pattern yields the smallest peak-to-peak oscilloscope pattern. Because the nature of the nonlinearity was defined ahead of time, the gain without the 1% flattening is known. In most problems the error sources are unknown and no way exists to find the "correct answer." The only choice available is to rotate the Lissajous pattern until the peak-to-peak signal is minimum and use that criterion to define gain. The remaining pattern defines an associated nonlinearity figure. It is very important to note that the 1% error in

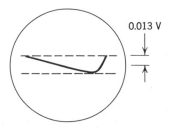

Figure 8.8 A rotated error display from Figure 8.7.

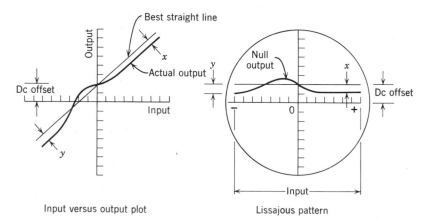

Figure 8.9 A comparison of best-straight-line errors and a Lissajous pattern.

a specific region of the output voltage is measured as an error of only 0.26% when related to the full output voltage pattern. The question raised here is this: Should the gain be defined for full-scale signal processes or over specific regions of the output? The latter definition is possible but seldom used. Gain is usually defined as a *full-scale* phenomenon, since the user does not know beforehand what operating region of the amplifier will be in use. Gain is usually defined for a no-load output condition. Gain variations caused by load changes are specified by the output impedance.

8.8 GAIN MEASURE—NONLINEARITY MEASURE OF DC AMPLIFIERS

The plotted slope of input signal versus output signal on an oscilloscope or on graph paper provides a means for measuring gain. Small departures from a best straight line are not easily observed, and for this reason the null technique just described is very useful. Such departures are called linearity errors. Figure 8.9 shows a typical plot of input versus output voltage with a dc output offset voltage. The errors have been exaggerated so that their relation to a null pattern is apparent. Several ideas are illustrated here.

1. The straight line used passes through the output value for *zero input signal*.

2. Only *one* straight line was used through the entire plus and minus response region.

3. The end or maximum points are not necessarily on the line.

4. The peak positive and negative errors are not necessarily equal. One polarity of error may even be nonexistent.

To illustrate these points further Figure 8.10 shows several typical error patterns in null form with a proper interpretation of peak-to-peak nonlinearity and a best null gain setting.

The foregoing discussion points up clearly that gain and nonlinearity measure go hand-in-hand. Once the gain has been properly measured, the nonlinearity figure is also available. With this background, gain and nonlinearity can be defined.

Definition. *Gain* is defined as the slope of the best full-scale straight line fitting the response values passing through zero input, not necessarily through the end points in the absence of phase shift. Best is defined as minimum peak error.

Definition. *Nonlinearity* is the largest peak-error departure from the best gain slope line defined above expressed as a percentage of peak output signal. Nonlinearity is usually stated as a ± value even though one polarity of error may be absent.

As stated before, gain as defined above implies a full-scale signal process. A 0.1% full-scale gain error could be a 1% gain error at 10% of full scale. If accuracy must be related to a fraction of signal level, the smallest full scale is a critical parameter. Some nonlinearity patterns have large gain errors referred to limited full scale. A pattern with end-point nonlinearity has nearly constant gain as the full scale is reduced.

Accurate gain measures usually involve the noise content present in the output of the device being monitored. Null techniques permit the eye to observe the envelope of the error if the noise cannot be removed by filtering. If the signal being processed is 10 Hz, a 100-Hz filter (e.g., a capacitor placed at the summing junction) can be added to eliminate noise above 100 Hz to facilitate the measure. If the capacitor modifies the error envelope, the filter is too severe.

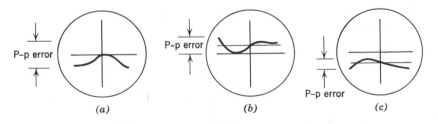

(a)   *(b)*   *(c)*

Figure 8.10 Several linearity error patterns.

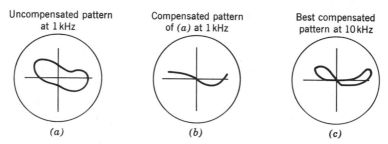

Figure 8.11   Typical Lissajous linearity error phase patterns.

8.9   PHASE SHIFT AND THE NULL LISSAJOUS PATTERN

If the amplifier signals are sinusoids and they are shifted in phase by the amplifier, the null pattern shown above becomes elliptical. If a linearity error exists, the pattern is semielliptical or simply a two-dimensional closed figure. If the amplifier has hysteresis, the pattern may have several closures.

Small amounts of phase shift can be balanced out by added trimming capacitances across the summing resistors, as in Figure 8.1. For amplifiers with excellent linearity, 45° phase compensations are possible. The phase correction produces a gain error factor which is calculable by (1) in Section 8.3. When linearity errors are present, they constitute harmonic frequency content not present in the signal generator. Phase compensation changes the gain to these higher frequencies and makes gain or phase measurements difficult to interpret.

If the gain is to be measured to an accuracy of 0.1%, the null errors for a 10-V signal should be reduced below 5 mV. A phase error of 1° causes a reactive component of 1.7%, which appears as 170 mV in the amplifier output (85 mV at the null point). It is apparent that very small phase errors are detectable and must be resolved before accurate gain measures can be made by this technique.

The null technique of measuring gain thus involves a phase correction to obtain a closed Lissajous pattern. If a nonlinearity is present, the closed pattern will be frequency-dependent. Above a certain frequency, no phase adjustment can be found that closes the pattern. A few typical patterns are shown in Figure 8.11.

8.10   SQUARE-WAVE GAIN TESTS

Sinusoidal null techniques require constant attention to phase compensation. Square-wave signals provide a gain observation technique that

is not dependent on reactive balancing. The technique requires interpretation, but it is a powerful measuring tool. Each transition of the
square wave presents a step-function input to the dc amplifier being
tested. If the period of the square wave is adequate, the amplifier will
settle to its final value before another transition occurs. The settling
phenomenon here is assumed to involve only energy storage which defines
the high frequency response of the amplifier; ac amplifiers with low-
frequency time constants are discussed later.

The step response of an amplifier is a measure of its stability. Excessive overshoot or ringing implies the possibility of instability under
slightly different operating conditions. The settling period is an important
measure in amplifier performance, but for gain measurements the time
after settling is of interest.

Gain measurements using square waves do not use a Lissajous pattern.
Instead the null pattern is observed on a normal linear time base sweep.
Figure 8.12 shows a typical square-wave response and the corresponding
null pattern. Pattern *b* defines a proper null. Pattern *c* shows how the
signal appears slightly off null with the horizontal line segments mis-
aligned. The only activity on the *y* axis should occur during output

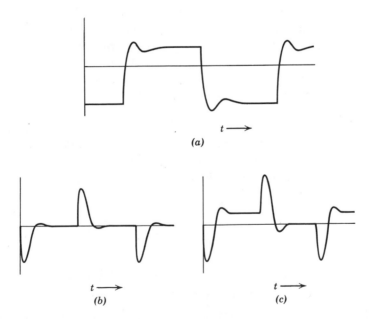

Figure 8.12 Square-wave null patterns: (*a*) actual output, (*b*) proper null,
(*c*) off null.

transitions. This technique obviously allows a measure of settling time to a specified accuracy.

The square-wave signal can be modified in three ways, (1) amplitude, (2) period, and (3) polarity. To ensure that the gain measured in Figure 8.12*b* is meaningful the result should be observed as a function of signal level and polarity. Any variations that occur are an indication of non-linearity. The gain measured at any one square-wave level, assuming a unipolar signal, is a measure of a single point along the sinusoidal linearity pattern. For rapid measure of gain accuracy minor linearity differences can often be dismissed; for example, if gain is measured to an accuracy of ±0.25% and the nonlinearity is 0.05%, a single-point square-wave test of gain is quite adequate.

Square waves measure gain over a band of frequencies rather than at a specific frequency. The longer the square-wave period, the lower the included frequency becomes. It is not sufficient to observe just low-frequency square waves as this procedure obscures high-frequency observations; for example, a 100-kHz bandwidth amplifier can be tested using 5-kHz square waves and low-frequency response can be monitored using 5-Hz square waves. With 5-Hz signals, the high-frequency response occurs in 0.1% of the duty cycle—a difficult viewing problem. Observing a 5-kHz square-wave response yields no information on gains below 5 kHz.

If the gain varies as a function of frequency, the horizontal segments of Figure 8.12 will not stay aligned as the square-wave period is varied. This response pattern is shown in Figure 8.13. Gain variation with frequency can occur if the open-loop gain versus frequency characteristics of the amplifier are not proper. Gain differences versus frequency of 0.1% are easily measured by this method.

In systems in which crossover phenomenon occurs, the gain either side of crossover can be very different. If the crossover is at 2 Hz, square-wave test frequencies below 0.1 Hz are recommended. It is possible to use a battery and a hand-held switch to provide these long period signals.

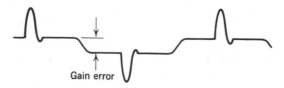

Figure 8.13 A frequency-dependent gain pattern.

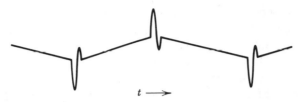

Figure 8.14 A square-wave null pattern for ac amplifiers.

8.11 GAIN MEASURE ON AC AMPLIFIERS WITH SQUARE WAVES

Ac amplifiers have a leading phase shift at low frequencies and a lagging phase shift at high frequencies. Zero phase shift occurs only at midband. In this region the phase angle changes with frequency inversely proportional to bandwidth. With phase compensation, sinusoidal null techniques are available over a limited frequency range. When square waves are used, the sag phenomenon associated with an ac coupled device is present, and this phenomenon is not simply canceled for a null observation. If the phenomenon is ignored and the alignment of two neighboring responses is considered a null, the null technique does provide gain information. The pattern is shown in Figure 8.14. If the alignment remains when the square-wave duty cycle is varied, the gain is constant with frequency. If the alignment is independent of signal level, the amplifier is linear. With ac devices, a measure of point-to-point linearity is very difficult to make and this technique yields a qualitative measure.

8.12 NOTES ON SQUARE-WAVE TESTING

The leading edge of a good square wave should have a rise time much shorter than the device being tested. If the amplifier is overloaded by the leading edge, its recovery processes should be considered. Some devices with limited slewing rate (see Section 8.17) have excellent recovery, whereas others do not. Some input-signal filtering may be necessary to keep a slewing problem from masking other parameters.

The full leading-edge signal is presented to the observation instrument immediately after each square-wave transition. In resistive summing back-to-back silicon diodes can be used to clip the null signals so they do not exceed 0.4 V. Since most null measurements are made at millivolt levels, the 0.4-V limit does not modify the measure. Some oscilloscopes are overloaded by 10-V microsecond pulses (1000 times overscale) at each square-wave transition, and it is important that their recovery processes are not confused with the amplifier's response.

If the oscilloscope is used differentially on a positive-gain amplifier, diode clipping should not provide a feedback path that can modify the amplifier's response. Resistance added to the points of observation will isolate the diode clipping and eliminate this possibility.

Noise present in the nulling operation is not usually a problem unless the noise level differs between adjacent pattern segments. The eye can usually line the segments up to the accuracy required to obtain a gain measure.

8.13 INSTRUMENT METERING OF GAIN

Gains measured with rms or average detecting devices are subject to a variety of errors. Meaningful measurements are possible and the following procedures should be followed. It is preferable that the same meter be used to measure the input- and output-signal level. If possible, external attenuators should be used to permit the instrument to be read at a single scale point at or near full scale. Gains can then be read as a function of resistor values, and the observation is limited by the quality and type of instrument involved and by the resolution provided. A panel meter can provide from $\frac{1}{4}$ to $\frac{1}{2}\%$ resolution. If a digital readout is presented, the reading will have greater resolution. Accuracy and resolution are often quite different. Greater resolution provides accuracy if the metering instrument is capable of accuracy; rms or averaging gain measure is meaningful only if the signals being measured are known to be free from distortion and noise. Freedom from distortion can be verified by using a low-distortion sinusoidal source and observing the resulting signal with a distortion analyzer. Freedom from noise contribution can be made by reducing the source-signal level to zero and noting the magnitude of any residual readings.

Distortion if present has no specific relationship to nonlinearity. A distortion figure of 0.1% rms can be composed of several harmonics varying in amplitude and phase. Depending on how these parameters are distributed, the nonlinearity figure can vary significantly.

The process of voltage averaging or rms averaging obscures the relationship between a signal and its harmonics. If the signal is free from harmonic content, then peak, rms, and average measure are all acceptable methods of measuring gains. If null techniques are not possible because of phase shift or other instrumentation limitations, then gain defined on an rms, average, or peak basis must be used—but qualified.

The gain definitions given in Section 8.8 assume a one-to-one relationship between input and output points after phase correction. Departures from this character are recognizable in null detection, but in any averag-

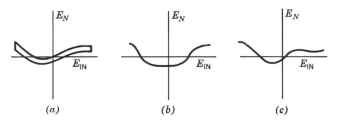

Figure 8.15 A few troublesome null plots: (*a*) bistable operation, (*b*) dynamic zero shifts, (*c*) shifts in linearity pattern.

ing process a departure is easily obscured. This class of signal disturbance can be caused by slewing limitations, distortion, or regional instability. Figure 8.15 shows a few troublesome null plots that cannot be sensed by a single average-gain observation.

A thorough gain analysis in the presence of significant phase shift is difficult. Special phase-shift networks can be added to overcome the limitation of the phase compensation circuit of Figure 8.1. Gain measure by instrument metering can be correct, but there are many possible pitfalls. Since gain and nonlinearity are interrelated, a measure of one without a knowledge of the other has questionable value.

8.14   TRANSIENT-RESPONSE TESTING

All feedback systems should be designed to provide a stable response to transient excitation. An evaluation of system performance should always include an observation of the transient response. The response can be nonlinear and a function of operating parameters. A meaningful evaluation should take these matters into consideration. To an experienced observer the response can also indicate possible hidden or obscured difficulties.

Linear operation must be defined before nonlinear behavior can be appreciated. Figure 8.16 shows the linear step response of an amplifier. The time of peak response is independent of signal amplitude. The signal reaches 1% of final value in a time period independent of signal level. If each response were observed with an oscilloscope set to yield the same vertical deflection, the patterns would be identical.

Stability testing of an amplifier should always employ signals that are small enough to operate the amplifier in a linear manner. The dynamic behavior in a nonlinear region can appear better or worse depending on circumstance. This does not exclude the possibility that small signal performance can vary as a function of operating parameters.

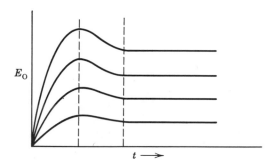

Figure 8.16 Linear step-response behavior.

Stability testing should include capacitive and resistive amplifier load variations as a function of gain setting and of input line unbalance. A few quick observations will indicate whether a complete search through all possible parameter combinations is necessary. In capacitive loading, small values should be tried (50 to 1000 pF) as well as larger values. Often a device is unstable for a 100-pF load but stable at 1000 pF.

8.15   FREQUENCY AMPLITUDE RESPONSE VERSUS TRANSIENT RESPONSE

The ideal transient response has less than 7% overshoot with little observable undershoot. In some applications a 1% overshoot figure is all that can be permitted. Frequency responses with minimum overshoot have a linear phase response. This can be explained as follows. The higher frequency content making up the step function is equally delayed in time to make up the response. Equal time delay for all frequencies is equivalent to phase shift proportional to frequency, that is, linear phase. The amplitude response of an amplifier with small transient overshoot has a very rounded character at the response knee. An optimally flat frequency versus amplitude response with low transient overshoot is *not* possible.

In a design that is marginal in frequency-versus-amplitude response, flat response can be obtained at the expense of transient overshoot. Typical response patterns are shown in Figure 8.17. The greater the peak in amplitude response, the more prominent the transient overshoot. Responses with more than one dominant overshoot are usually not acceptable.

Settling time is obviously improved if the rounded response ② is used. To measure the settling phenomeron to better than 1% a null technique

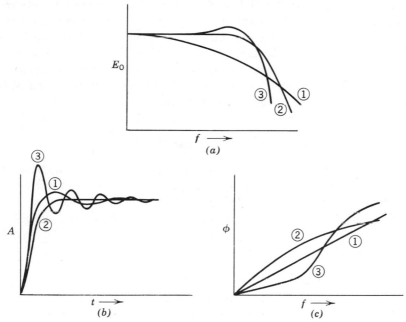

Figure 8.17 Transient response versus frequency response.

is recommended. Oscilloscope observations with zero suppression can also
be used. The discussion in Section 8.10 is applicable.

Rise time can be defined in a variety of ways. One definition is the
time from 10 to 90% of final value. If the overshoot is less than 10%,
this definition is usable. Direct oscilloscope measurements are usually
acceptable.

8.16 UNSTABLE TRANSIENT-RESPONSE CHARACTER

The principal overshoot at the output of an amplifier is not a full
indication of the margin of stability. Figure 8.18 shows a few response

Figure 8.18 Marginal transient-response conditions.

patterns that are suggestive of trouble. External pickup can sometimes present similar patterns. If parameter changes affect the repetition rate of the minor variations, the problem is not pickup. Patterns such as those in Figure 8.18 arise because of cancellation effects or because one amplifier's response is attenuated by another. If external passive isolating inductors or filters are used and are improperly damped, they can also produce patterns similar to those of Figure 8.18.

8.17   SLEWING-RATE LIMITATIONS

Slewing rate is simply the volts-per-second capability of an instrument or amplifier. A 20-kHz signal may be practical at a 1.0-V output level, but the amplifier may be incapable of a 10-V output swing at this same frequency. When a slewing-rate limit is reached, the amplifier responds in a nonlinear manner until it "catches up" with the signal. The amplifier then must recover from any accumulated overload and return to normal operation.

Consider an amplifier with a $+100$- and -200-mA output current limit loaded by a 0.1-μF capacitor. The current i in a capacitor C for a sinusoidal voltage $e = E_p \sin 2\pi ft$ is $i = c \, de/dt$ where f is frequency in Hertz and E_p is peak voltage. If $E_p = 10$ V and $f = 16$ kHz, the peak current is just 100 mA. The output voltage waveform for a 20-kHz signal is shown in Figure 8.19. The maximum slewing rate here is 10^6 V/sec and is defined by the ratio of current to capacitance. A 20-kHz 10-V sinusoid requires the voltage change at a maximum rate of 1.25×10^6 V/sec.

The signal shown in Figure 8.19 has a straight-line segment over a part of the waveform. The straight-line response is typical of an output-stage slewing-rate limit process. If the slewing limit occurs at an internal stage in an amplifier, the response may be modified. Feedback, if available, tends to correct the waveform leaving the effect somewhat more difficult to sense.

Figure 8.19   Output voltage waveform for a slew-rate limited amplifier.

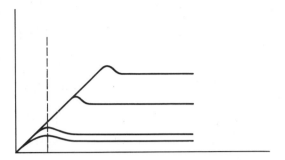

Figure 8.20   Slew-rate limited settling phenomenon.

The waveform in Figure 8.19 indicates signal rectification with the average voltage slightly negative. If the rectification process occurs at an internal amplifier point, the effect is a change in internal operating point. Loop gain tends to cancel the injected offset. The error can be sensed as a change in the dc output level even though the sinusoidal waveform may be preserved by feedback. The presence of a dc offset that varies with signal level is a strong indicator of slew-rate limiting.

Slewing-rate demands are greatest for a fast-rise square-wave input signal. Here settling time becomes amplitude-dependent. Figure 8.20 shows a typical settling pattern which should be compared with the linear behavior of Figure 8.16.

Slewing-rate limitations can be made symmetrical so that no net dc offset results. The only external manifestation might be in terms of a frequency-versus-amplitude response that is amplitude-dependent. This character is shown in Figure 8.21.

Slew-rate limiting in ac amplifiers that would cause a dc shift in a dc amplifier can be obscured by decoupling elements. When a slewing-rate limitation occurs, the response after signal removal may have an

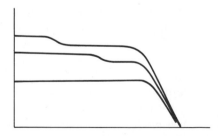

Figure 8.21   A slew-rate limitation effect obscured by feedback.

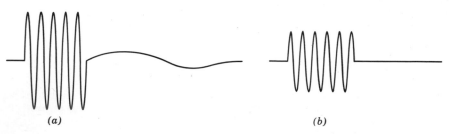

Figure 8.22   Ac amplifier slew-rate recovery.

observable recovery transient. Figure 8.22 shows a problem response and a smaller signal with proper response. Note that the sinusoid is high frequency and a one-half cycle rectification could not cause the indicated response. It is important to verify that the recovery process is amplitude-dependent before interpreting the recovery as slew-rate related.

8.18   OUTPUT IMPEDANCE

Output impedance is defined as the ratio of change in output voltage to a change in output current. This definition excludes any measure of static offset voltage. For output impedances below a few ohms, a null measure of gain change with and without load can be used to measure output impedance. The nonlinearity pattern also indicates that voltage limiting does not occur at full current. If a square-wave test is used, both full-scale polarities of signal should be used. The percentage of change in gain times the load value used is the output impedance.

The path of output current must be known to obtain a valid impedance measure. If the load is placed directly on the output terminals, excluding the connector, the resistance of the connector will be eliminated. The most damaging current path would include any part of the input signal conductor. If both generator and meter are grounded, a resultant ground loop would make an output impedance measure meaningless.

The output impedance of a dc amplifier is inductive at high frequencies because the low output impedance is obtained from feedback and the loop gain is intentionally reduced at high frequencies to provide amplifier stability. It is also standard practice to place a small RL parasitic suppression circuit in series with the output terminal which adds to the inductance. This impedance modifies the gain in the presence of a load. A measure of impedance change with frequency is a measure of the inductance.

The output impedance is obviously not as simple as a single inductor, but the first-order inductance at the upper response frequency can be approximated by a few measurements. Null measurements should be made at as high a frequency as practical to obtain any observable gain change effects.

A dynamic method of measuring output impedance involves pumping current into the output terminals from an external generator. This measure excludes connecting a signal to the input terminals. The input should be connected to a source of zero signal through a correct source impedance. A resistor should be placed in series with the signal generator so that the current flowing into the amplifier output is measurable. If the output impedance is low, the impedance is the ratio of amplifier output terminal voltage to signal voltage times the series resistor used. The amplifier output signal should be measured ahead of any connectors to eliminate their resistance value.

8.19   INPUT IMPEDANCE

Input impedance is defined on a difference basis, that is, the ratio of change in input-signal voltage to change in input current. Source current is a separate phenomenon and, if included in the measure, can yield misleading results. In many devices the input impedance at low frequencies is in excess of 1000 MΩ. Since 2.7 pF has a reactance of 1000 MΩ at 60 Hz, it is easy to see why these very high values are difficult to measure above a few hundred hertz.

Source current can be measured by noting the change in output offset as a function of source resistance; for example, if a gain 1000 amplifier has a 1-mV offset change for a source-resistance change from 0 to 1000 Ω, the input current is 1 μV in 1000 Ω or 1 nA. Changes of 1 μV can arise from thermocouple effects, and for this reason the techniques used to effect the resistance change must be isothermal.

High input impedance does not imply an ability to accomodate high source impedance. In many designs the high impedance is present to assure CMR, and source impedances above 1000 Ω are not recommended. One method of measuring input impedance with a source resistance of 1000 Ω involves the square-wave null technique of Section 8.10. The gain at a low repetition frequency, say 10 Hz, is measured with and without a 1000-Ω source. The gain-change percentage divided into 1000 is the input impedance. A 0.1% change corresponds to 100 MΩ. If performance parameters are not disturbed, a 10,000-Ω change yields a larger gain change.

An input capacitance of 100 pF is not uncommon. If the input impedance is high by virtue of feedback at a frequency at which shaping networks have reduced the loop gain, the input impedance is also reduced. Shielded input cable shunts the intrinsic active capacitance; 40 to 60 pF in cable capacitance is typical in test runs. Note that at 10 Hz, 10 pF is 1000 MΩ.

The above reasoning indicates that any measure of input impedance must accommodate this input reactance problem. If resistance changes are used, they should *not* be made at the generator end of an input cable but rather at the amplifier input terminals. If this is not done, the measurement may be dominated by the cable capacitance.

To be useful an input-impedance measure should include a resistive term and a capacitive term. A 6-dB slope per octave on input impedance defines the reactive term. Since active devices rarely exhibit such a "clean" character, the measure will be an approximation, usually dependent on amplifier-gain settings.

The input impedance of an operational amplifier is its input resistor and does not require measuring. At high frequencies the shunt capacitance across the resistor reduces the input impedance. Input cables usually dominate this effect however.

Output-signal currents can raise or lower the input impedance if these currents flow in an input-signal conductor. If the input impedance changes as a function of output load resistance, the measurement should be suspect.

Impedances which define CMR are best measured by testing CMR. If the CMR impedance point is at the input terminals, the CMR *in the presence* of a known line unbalance is a straightforward method of measuring input impedance. A CMR of 10^6 at 60 Hz for a 1000-Ω unbalance implies an impedance level of 1000 MΩ. If CMR falls off at 6 dB/octave, this 1000 MΩ is shunted by a capacitance. If the CMR is nearly constant with frequency, it usually implies the 1000 MΩ is resistive.

Negative input impedances can exist and should be viewed cautiously. Any null measure should show a decreasing impedance value with increasing frequency.

In slew-limiting processes in which potentiometric feedback is used the input impedance can drop to very low values over parts of a cycle. This phenomenon is a serious one particularly in multiplexers when large signal changes are encountered. If the input impedance is a function of signal level, slew limiting is probably occurring. Square-wave testing provides the most rapid input change possible, and a null pattern can indicate very clearly the presence of this type of phenomenon.

Capacitive input sources block the flow of source current, but they also can "block up" the amplifier. (See Section 7.15 for a discussion on source-current path requirements.) In most dc amplifiers a resistive path across the signal lines, as well as a ground return through the CM source, is required.

8.20   OUTPUT VOLTAGE AND CURRENT

Voltage and current specifications do not always state whether maximum values must occur together. A 10-V–100-mA specification can be tested together by applying Ohm's law and using a 100-Ω load.

Maximum values should always be observed at low line voltage as this is where output stages are often limited in capability. Tests should be made at very low frequencies to ensure that all coupling mechanisms are adequate. Metering at low frequencies is best handled on a direct-coupled oscilloscope on which the entire waveform is visible. To test very low frequency performance square waves are recommended. A 1-Hz square wave at full scale that shows a lack of sag, hum, or instability implies excellent low-frequency performance. Both polarities of signal should be tried.

High-frequency performance can appear adequate but slewing-rate limitations may still be present. The waveform at the highest frequency should be free from distortion, but this is not necessarily a proof that slew limiting is absent. The frequency response curves of Figure 8.21 show one slew-rate situation. If filters eliminate the distortion after the limiting point, the direct signal observation may be distortion-free and thus misleading.

Some designs are marginal at a fixed dc output current level although they are capable of supplying the current on an intermittent basis. The limitation is often in the overheating of transistors, resistors, transformers, or just heating in general. The usual practice is to design class B stages and use feedback to remove any crossover phenomenon. At full output current, the stages are no longer drawing a quiescent current and are dissipating maximum power. A good half-hour run at full static dc output power will show up any deficiencies. Full output power for both output polarities should be considered. In some designs a worst-case output signal exists and should be used. High line voltage and high ambient temperatures also aggravate the situation. Testing under these worst conditions may seem unrealistic, but instruments running unattended for hours often get just this treatment. Some designs are capable of passing the above tests without special ventilation, whereas others require air passage supplied by a blower system. If blowers are supplied, the test should include their use.

All performance tests should be made with near-full-scale signals.

It is incorrect to test a gain 1000 unit at 10 V, reduce the gain to 100, and perform the same tests at 1 V. To obtain meaningful results, signal levels must be increased as gain is reduced.

The maximum current output of an instrument should occur near full-scale output voltage. If low voltages are used as full scale, the accuracy will be low. Consider a 0.1-Ω output impedance and a 10-Ω load. The current error will be 1% because of the 0.1-Ω impedance. The voltage across the 10 Ω is thus 1% low. The full-scale output voltage for 100 mA is 1 V. If lower voltages are used as full scale, the errors are proportionately greater.

All specification errors are related to full-scale output voltage unless specifically qualified otherwise. In general, any errors referred to the output are increased directly as the full-scale output level is defined lower.

8.21   DRIFT AND TEMPERATURE COEFFICIENT

Three components of drift are usually present:

1. Drift referred to the input (RTI)
2. Drift referred to the output (RTO)
3. Drift referred to input-source current

Tests which separate these variables are possible, but the necessity for this exercise varies depending on the design. If RTO drift is small compared with RTI drift, the latter is usually the only parameter measured. Engineering evaluations require a full set of tests.

Drift RTI is usually measured at full gain to place microvolt effects in the millivolt range. Continuous output recording rather than sampling provides the best measure. Most recorders are limited to below 10 Hz bandwidth, and this upper limit of response is desirable in recording drift tendencies.

Many designs require a warmup period before a recording is taken. This time period permits most thermals to reach equilibrium. The total time of observation varies depending on the test intent. A good measure of response can be made in 20 min although 24-hr runs are not uncommon. Note that drift qualities can appear in 24 hr that are not visible in a 20-min run.

Repeatable zeros are not always inherent in a design. Each time a unit is turned on, it may take a new dc set. The differences may be only microvolts, but this can be a problem in some applications. Long drift runs can sometimes yield an indication of this phenomenon. At some time during the run, the unit may change zero in a stepwise manner.

The phenomenon is almost always a function of the semiconductors used. With present computer techniques for monitoring zero values, such an error can be accommodated. Also, as semiconductors improve, the phenomenon is less likely to occur.

Drift measures should be made in the final environment; that is, in a cabinet or in a rack mount mounted above other equipment. This treatment ensures that the thermal conditions and air conditions are typical. It is not uncommon for a device to meet its specifications on a bench, free from cabinetry, and then fail miserably when it is rack mounted and when air is supplied from an internal blower.

Drift measures made with shorted input lines ensure that source current (pumpout) effects are eliminated. Since this current varies with temperature, a source impedance will modify the drift character as now two components of zero error are included. In designs with temperature-regulated input or mechanical choppers the current can be near zero or if present, very stable. One nA in 1000 Ω is 1 μV. If the current changes 0.2 nA/°C, a 20°C change produces 4 μV of drift. For fixed temperature operation the effects of source current can be neglected.

Drift RTO can be made by checking an amplifier at low gains. That component of error not attenuated by gain change is said to be RTO. Typical figures might be ± 0.1 mV at constant temperature in 40 hr. Drift RTO is not present if one overall feedback loop surrounds the amplifier.

Much of the drift in amplifiers occurs because of temperature changes that take place over very long periods of time. It is thus not meaningful to measure drift independent of temperature. If controlled temperature changes can be supplied by a controlled chamber, drift versus temperature (TC) can be separately measured. The thermal mass must be kept reasonably small, and it is standard practice to perform these tests on an amplifier without its usual cabinetry. Note that the input leads should be entirely within the chamber.

Environmental chambers are usually very turbulent, and if the design is sensitive to air motion, a measure of TC requires taking special precautions. One technique is to cover sensitive areas with a blanket. As a result the temperature in critical areas change slowly, unaffected by short bursts of air.

As the temperature changes, the output zero responds slowly. The curve of zero error versus temperature is rarely a straight line. In general it is a curve representing the balancing of several exponential functions. By reading a specification, it is not possible to determine whether a temperature-coefficient statement is an average value or a maximum slope value. The latter is more meaningful as it relates to worst-case

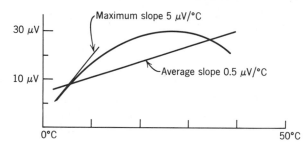

Figure 8.23 A temperature-coefficient curve showing average slope.

operating conditions. The former value is a smaller number as it permits averaging over a larger temperature region. From a reading of the specification, no information is contained about the worst that can be expected. Typical temperature curves are shown in Figure 8.23.

A drift specification should always be accompanied by a temperature-coefficient specification. If not, the test conditions and temperature control required are undefined, and drift tests are not comparable between users. A tight control of temperature can yield test results that can be compared with a specification but this is often hard to do. Specifically it is not a practical constraint to apply, for the user wants to know what to expect in his operating environment.

The full temperature specification thus has six parts:

1. Drift RTI
2. TC RTI
3. Drift RTO
4. TC RTO
5. Source current
6. Source current TC

Designs with small temperature-controlled regions (ovens) control both the TC RTI and the source-current TC. Designs without ovens can compensate for base-emitter effects to obtain a low TC, but the source current may then vary with temperature. Operation with a maximum source impedance should be observed to measure the worst-case drift condition.

Some designs are very sensitive to ambient air flow. They are well within their specifications *on the average,* but if a door is opened in a room, the output is affected for 10 min until the thermal transient has died down. The thermal isolation of input stages is a key design parameter for a good amplifier. In specification evaluation the possibility of poor design must be considered. Needless to say, it is not a quality approach—it is however clever specification dodging.

Low-frequency noise below 1 Hz is not easily measured except in drift runs such as those above. A drift specification is a dc measure, and strictly speaking it is an ac signal with a period equal to the time of observation. Average values taken over long periods of time are sometimes used in obtaining low drift numbers. A more useful description is peak-to-peak error in a stated number of hours for a fixed bandwidth. A drift figure of ± 2 μV peak-to-peak in 3 Hz bandwidth in 40 hr would be a proper and useful statement of drift. Note that a dc parameter is described in terms of ac properties.

8.22   COMMON-MODE REJECTION TESTING

The rejection of ground differences of potential can be checked by connecting a signal generator from the output zero-reference conductor to the input zero-reference conductor. This assumes that a short circuit does not exist between conductors through a ground path. Leakage paths of 1000 MΩ or a few picofarads can reduce the CMR; therefore the signal circuits must be very well shielded. The test has meaning only if a line unbalance is present to sense the flow of common-mode current. A test circuit is shown in Figure 8.24 with a 1000-Ω potentiometer to provide source unbalance in either input line. The CMR ratio is 20 log (E_{CM}/E_0) $\times$ gain; CMR tests are often made at or near 60 Hz. Note that the shielding around the potentiometer must be tight. It is important to observe the CMR ratio as a function of frequency. In some designs, CM slewing rate is a problem. Since unwanted CM signals are often "spikelike" or "pulselike" in character, the design should not overload under these conditions. If the design has been balanced specially to perform at 60 Hz, it may fail at higher or lower frequencies.

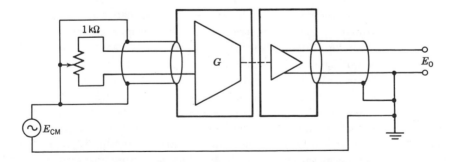

Figure 8.24 A typical common-mode test circuit.

In designs with postisolators the CMR is constant with gain but falls off proportional to frequency. In designs in which CMR is obtained actively, the CMR ratio may stay constant to several hundred cycles and fall off linearly with gain. (See discussions in Section 7.11.) In many active CMR designs, the CM signal level and input signal level combined cannot exceed say 10 V. Serious limitations are apt to occur at the lowest gains when input signals are large. CMR at zero frequency can be tested with the circuit in Figure 8.24. If heating occurs, the CM signal may cause an unbalance that will result in poor CMR and or excessive output zero shift.

Signal common-mode tests are performed when the rejection process is separated in the design. If the signal common mode is static, such as with a resistance bridge, this type of source can be simulated either by operating with a bridge present or by adding a dc source between the shield and the signal lines. Resistance bridges are balanced sources, and the CMR ratio should be very high under this condition. To test CMR ratio with line unbalance and with a resistance bridge present, the unbalance resistor can be added to the signal lines between the amplifier and the signal connection.

8.23 TEMPERATURE COMPENSATION AND ACCOMMODATION

Two procedures are available to reduce temperature effects: compensation circuits and ovens. Mechanical chopper circuits are not generally compensated for temperature as the chopper itself limits the performance.

Symmetry in a differential sense has been discussed in Sections 5.6 and 7.15. As temperatures change, the changes in transistor characteristics are nearly equal. Each transistor in a differential stage affects the output in an equal and opposite manner and the combined effects cancel. A second-order drift term is usually left. To correct for this effect a temperature-sensitive circuit can be added to inject correction currents as a function of temperature. The elements used can vary and might include thermistors, diodes, or transistors. By proper circuit planning the current can be added in either sense and in a varying magnitude. By use of shunting resistors, the slope of the injected current can be varied. A typical diode injection circuit is shown in Figure 8.25.

To adjust the compensation circuit properly, a temperature test must be made. The difficulty this poses is shown in Figure 8.23 in which an average specification is defined. The temperature coefficient over a narrow temperature region can be much greater than the average value over a larger temperature span. Note that the compensation current can also be added to the collector circuits.

Temperature compensation to accommodate base-emitter changes often

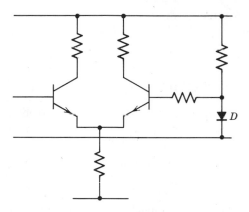

Figure 8.25 Diode temperature compensation.

neglects source-current variation. If the current is low enough, the temperature coefficient is unimportant. For high source impedance applications, the omission can be serious.

Temperature-sensitive circuits must be physically associated with the circuits they are compensating. If not, variations in ambient conditions will cause the compensation to be in error. It is usually necessary to mount compensating circuits and input stages together and then to insulate the region thermally so that thermal transients are attenuated if they occur. The design should not include heat-producing elements in the vicinity. The comments on ventilation in Section 8.21 are applicable.

Ovens containing the input stages can reduce the TC of source current as will base-emitter effects. The oven can include temperature-sensitive resistors, power-supply sensing elements, and feedback elements as well as input stages. The smallest ovens regulate the temperature of the substrate region only for the input transistor pair.

The larger the thermal mass of the oven, the larger the warmup time. Thermal substrate heaters reach equilibrium in a few seconds. An oven weighing 10 oz will take 10 to 20 min to stabilize. As of this writing, transistors on thermally regulated substrates are noisier than other available low-level stages and their application is somewhat limited.

Oven temperatures must be higher than the highest internal ambient, or regulation will not be possible. A good thermal regulator should produce a near-zero temperature gradient in the area of sensitive elements. If a gradient exists, it will vary as a function of the external ambient, and the resulting temperature differences will produce zero errors. Good oven regulation requires that heat flow out of the oven as well as in. Note that the elements in the oven do produce some heat. The heat flow out of the oven determines the temperature slewing rate. For a fast settling time it is desirable to have the slew rate in reasonable bounds.

8.24 LATCHUP AND MOTORBOATING

An overload phenomenon that can occur in feedback amplifiers is called "latchup." One form of the phenomenon occurs when an amplifier under some condition of input signal becomes inoperative with internal operating conditions badly disturbed. Usually the amplifier rests at one extreme of full output voltage. The immediate "fix" is to turn the power "on and off" and start operation all over again.

A latchup usually occurs when internal operating points change so that a gain stage with a normal polarity reversal is saturated and a nonreversing signal path still exists through or around this stage. The reversal in gain polarity changes normal negative feedback to positive feedback. This further guarantees that the offending stage will remain saturated. Positive feedback continues to support the latchup, and no normal way exists to return to proper operation.

It is easy to see that latchup conditions can be difficult to excite; for example, they can be marginal and occur only statistically in the same design. The proper conditions are related to internal operating points, specific stage gains, power-supply voltages, or certain gain and offset conditions.

Differential stages are offenders as they can reverse gain polarity depending on succeeding stage loading. A good design will ensure that the circuit loses gain in overload while retaining gain polarity. Sometimes a diode clamp on the positive-gain side will ensure this condition.

Positive feedback, latchup, and oscillation all go hand-in-hand. An amplifier in latchup may or may not be oscillating. Sometimes evidence of "near latchup" is visible by observing that the amplifier tends to "burst" or "tear" at extremes of signal level. Here, again, a slewing condition may have to be present to cause a sufficient change in internal operating point to force the effect to the surface.

"Motorboating" is a low-frequency phenomenon usually caused by a form of latchup. The amplifier overloads and changes gain polarity, and the resulting oscillation forces the amplifier to another extreme of operation. The recovery process brings the amplifier back to the latching region and the process is repeated. The appearance is one of low-frequency oscillation, but quite frequently it is a high-frequency-related phenomenon.

Motorboating requires long time constants, which are present in ac-coupled carrier amplifiers or when signal transformers are used. Motor-boating is not a usual phenomenon in purely direct-coupled devices unless heating is the affected time constant in the process.

Evaluation of amplifiers should always contain a plan to force latchup or motorboating. Low line voltage operation and overscale signals at all frequencies should be tried at all extremes of gain (feedback). More than one instrument should be evaluated to ensure a sampling. When differential amplifiers are evaluated, both signal- and ground-difference common-mode signals should be tried. High-frequency common-mode content can cause latchup where high-frequency normal mode signals may have no effect. The area is complex and requires ingenuity on the part of the evaluator to consider all meaningful operating extremes. The author has seen some forms of latchup go unnoticed for literally years because the right combinations of operation did not happen to be tested. This sort of find is very unnerving to the designer who after a year or so is "confident" of his design effort.

PROBLEMS

1. Use Figure 8.7. Sketch the patterns that result if the input signal is a triangle wave rather than a sine wave. Full scale is ±10 V.
2. Draw a null pattern if the amplifier under test has a 10% gain increase for positive signals.
3. What is the null gain measurement of Problem 2 if the negative gain is 100?
4. An amplifier has 1% second-harmonic distortion in phase with the signal. What is the nonlinearity error?
5. In Problem 2, recalculate the null gain for a full scale of 5 V.
6. In Figure 8.1 does a capacitor from ② to ⑦ modify the phase-shift measure determined by the trim adjustment?
7. How much compensation for amplifier phase shift causes a gain error measure of 0.1%? Will the error be high or low?
8. What is the maximum slewing rate of a 10-V-peak 100-kHz signal? How much peak current flows in a 100-pF capacitor with this voltage impressed across its terminals?
9. If the signal in Problem 8 has a slew-rate limit of 5×10^6 V/sec, graphically draw the resulting waveform and estimate the percentage of reduction in peak-to-peak voltage.
10. In Problem 9 assume that the slew-rate limit occurs in only one direction. Estimate the dc offset that results.
11. Draw a block diagram of an amplifier that might exhibit the qualities of Problem 10.
12. An amplifier has a 1-Ω output Z at 1000 Hz and 3 Ω at 100 kHz. What is the approximate output inductance.
13. An amplifier measures gain 50 with a 1-kΩ source and 49.92 from a 10-kΩ source. What is the input Z? At 10 kHz the gain measures 49.00 from a 10-kΩ source. What is the input capacitance approximately? Assume the amplifier has a bandwidth to 50 kHz.

9

Active Devices

9.1 INTRODUCTION

Operational and potentiometric feedback techniques can be used to design active filters. Such filters provide a low output impedance so that the response characteristics are not load-dependent. The use of amplifiers eliminates the need for inductors as the desired transfer characteristics can be obtained by using resistors and capacitors. The filters that result can have a high input impedance and gains greater than unity and can be simply modified by relay or switch selection. These qualities make active filters more flexible than corresponding passive devices.

Filter transfer characteristics are defined in terms of natural frequencies. If the filter is of high order, several cascaded active circuits may be required to obtain the desired characteristics. A section can be designed to accommodate a high-pass or low-pass characteristic with given natural frequency and damping or a simple RC or LC characteristic. In terms of the complex frequency variable s, each section can be designed to have pairs of complex left-half plane poles in the s plane, poles on the negative real axis, or zeros at the origin. Operational feedback and potentiometric feedback provide negative and positive gain, respectively. Input impedance, input slewing rate, or gain polarity defines the selection of feedback type.[1]

---

[1] The material that follows discusses a few standard forms and is not intended to be a complete treatment of active filters. The pole locations for various standard filters are given in the literature.

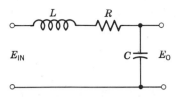

Figure 9.1   A low-pass section.

9.2   PASSIVE-VERSUS-ACTIVE CIRCUITS

A passive low-pass second-order section is shown in Figure 9.1. The transfer function in terms of the complex frequency operator s is

$$\frac{E_O}{E_{IN}} = \frac{1/LC}{s^2 + sR/L + 1/LC} \tag{1}$$

The natural frequency, in which E_O and E_{IN} have a 90° phase relationship, occurs where

$$s = j\omega_0 = j/\sqrt{LC} \text{ or } \omega_0 = 1/\sqrt{LC} \tag{2}$$

The coefficient R/L can be written in terms of the damping ratio ζ and the natural frequency ω or

$$\frac{R}{L} = 2\zeta\omega_0 \quad \text{where} \quad 2\zeta = \frac{R}{\sqrt{L/C}} \tag{3}$$

If $\zeta < 1$, the roots of the denominator are complex, and the transfer characteristic exhibits a step-function overshoot. The damping ratio where $\zeta = 1$ is called the critical damping.

An active filter that is equivalent to the passive filter is shown in Figure 9.2. The signal at ① is E_O/G and the current I in R_2 and C_2 is

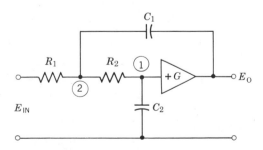

Figure 9.2   An active filter section.

$E_0 s C_2/G$. The voltage at ② is $E_0/G + IR_2 = E_0/G + R_2 E_0 s C_2/G$. Using Kirchhoff's law at node ② gives $(E_2 - E_{IN})/R_1 + E_0 s C_2/G + (E_2 - E_0) s C_1 = 0$. Solving for E_0/E_{IN} yields the transfer function

$$\frac{E_0}{E_{IN}} = \frac{G/R_1 R_2 C_1 C_2}{s^2 + s[1/R_2 C_1 + 1/R_1 C_1 + (1 - G)/R_2 C_2] + 1/R_1 R_2 C_1 C_2} \quad (4)$$

The coefficients in the denominator can be related to natural frequency and damping as in the passive case above. Thus $\omega_0 = (R_1 R_2 C_1 C_2)^{-\frac{1}{2}}$ and the middle coefficient equates to $2\zeta\omega_0$. Dividing by ω_0 gives

$$2\zeta = \left[\left(\frac{R_2}{R_1}\frac{C_2}{C_1}\right)^{\frac{1}{2}} + \left(\frac{R_1 C_1}{R_2 C_2}\right)^{\frac{1}{2}} + \left(\frac{R_2 C_2}{R_1 C_1}\right)^{\frac{1}{2}} (1 - G) \right] \quad (5)$$

Substituting $R_2 = a^2 R_1$ and $C_2 = b^2 C_1$ gives the damping ratio

$$2\zeta = \frac{b^2 + 1}{ab} + ab(1 - G) \quad (6)$$

By selecting only ratios of resistors and capacitors and a gain factor G, any damping ratio ζ is possible. If $G = 1$, the minimum damping factor occurs for $b = 1$. To obtain a damping factor of 0.5, $a^2 = 4$. Note the dc gain is G.

The complex pole locations are at the roots of the denominator polynomial. The denominator polynomial $p(s)$ is

$$s^2 + 2\zeta\omega_0 s + \omega_0^2 \quad (7)$$

The roots are at $s = -\alpha \pm j\beta$, where $\alpha = -\zeta\omega_0$ and $\beta = \omega_0 (1 - \zeta^2)^{\frac{1}{2}}$. The pole locations and the relationship to damping fraction are shown in Figure 9.3.

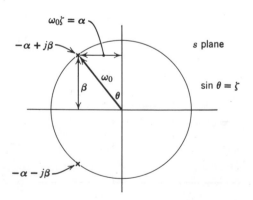

Figure 9.3   Pole locations in the *s* plane.

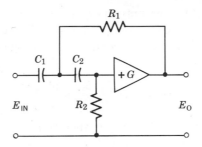

Figure 9.4 A high-pass filter section.

9.3   A HIGH-PASS FILTER

The high-pass equivalent to Figure 9.2 can be obtained by interchanging resistors and capacitors. Such a circuit is shown in Figure 9.4. The natural frequency (90° phase shift) occurs at $\omega_0 = (R_1 R_2 C_1 C_2)^{-\frac{1}{2}}$. Using the same element ratios a^2 and b^2 again gives the damping ratio, as $2\zeta = (b^2 + 1)/ab + (1 - G)ab$. This circuit has the transfer function

$$\frac{E_O}{E_{IN}} = \frac{Gs^2}{s^2 + [1/R_2C_2 + 1/R_2C_1 + (1/R_1C_1)(1 - G)]s + 1/R_1R_2C_1C_2} \tag{8}$$

with two zeros at the origin. The high-frequency gain is G.

9.4   LOW-PASS FILTERS WITH NEGATIVE GAIN

An active filter using a negative-gain amplifier is shown in Figure 9.5. The transfer function is

$$\frac{E_O}{E_{IN}} = \frac{1/R_2R_3C_1C_2}{s^2 + s(1/C_1R_1 + (1/C_1R_2 + 1/C_1R_3) + 1/C_1C_2R_2R_3} \tag{9}$$

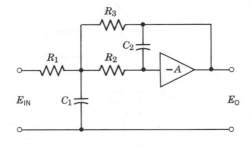

Figure 9.5 A negative-gain active filter.

The natural frequency occurs where $\omega_0 = (C_1 C_2 R_2 R_3)^{-\frac{1}{2}}$. Using the substitution $R_2^2 = a^2 R_1^2$, $C_2^2 = b^2 C_1^2$, $R_3^2 = G^2 R_1^2$ and noting that the coefficient of s in the denominator is equal to $2\zeta\omega_0$ where ζ is the fraction of critical damping yields

$$2\zeta = abG + \frac{bc}{a} + \frac{a^2}{G} \tag{10}$$

The gain at zero frequency is simply the value G. Frequency characteristics are not a function of gain A as long as sufficient loop gain is present.

The high-pass equivalent to Figure 9.5 is not practical. When capacitors and resistors are interchanged, capacitors must sum currents into the summing point. These currents are proportional to frequency, placing demands on the signal source as well as the amplifier output at high frequency.

A second useful negative-gain filter configuration is shown in Figure 9.6. The transfer function is

$$\frac{E_O}{E_{IN}} = \frac{R_3 R_4 C_3 [s + 1/R_3 C_3 + 1/R_4 C_3]}{R_1 R_2 C_1 (s + 1/R_1 C_1 + 1/R_2 C_1)[s^2 C_2 C_3 R_3 R_4 + s(C_2 R_4 + C_2 R_3) + 1]} \tag{11}$$

The zero of transmission in the numerator can be canceled by equating it to the simple pole in the denominator. Thus let $1/R_3 C_3 + 1/R_4 C_3 = 1/R_1 C_1 + 1/R_2 C_1$. By using this relationship, the transfer function becomes

$$\frac{E_O}{E_{IN}} = \frac{(R_3 + R_4)/C_2 C_3 R_3 R_4}{(R_1 + R_2)[s^2 + s/C_1 R_3) + 1/C_1 R_4 + 1/C_2 C_3 R_3 R_4]} \tag{12}$$

The natural frequency is

$$\omega_0 = (C_2 C_3 R_3 R_4)^{-\frac{1}{2}}$$

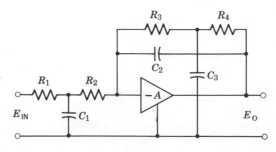

Figure 9.6 A negative-gain active filter with pole/zero cancellation.

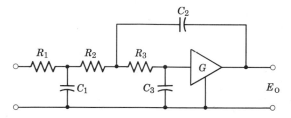

Figure 9.7   A third-order low-pass filter.

and the fraction of critical damping ζ becomes

$$2\zeta = \left(\frac{C_2}{C_1}\frac{R_4}{R_3}\right)^{\frac{1}{2}} + \left(\frac{C_2}{C_1}\frac{R_3}{R_4}\right)^{\frac{1}{2}} \tag{13}$$

The gain at zero frequency is

$$\frac{R_3 + R_4}{R_1 + R_2} \tag{14}$$

9.5   THIRD-ORDER FILTERS

The active devices shown so far provide two poles of transmission. If a third simple pole is required, it can be added by including an R_1C_1 element as in Figure 9.7. The transfer function is

$$\frac{E_O}{E_{IN}} = G\{R_1R_2R_3C_1C_2C_3s^3$$

$$+ s^2[R_2R_3C_2C_3 + R_1R_3C_2C_3 + R_1R_3C_1C_3 + R_1R_2C_1C_3$$
$$+ (1 - G)R_1R_2C_1C_2] + s[R_3C_3 + R_2C_3 + R_1C_3$$
$$+ (1 - G)(R_2C_2 + R_1C_2)] + 1\}^{-1} \tag{15}$$

A third-order low-pass filter with roots at $-\gamma$ and at $s = -\alpha \pm j\beta$ is

$$\frac{E_O}{E_{IN}} = \frac{G}{(s + \gamma)[(s + \gamma)^2 + \beta^2]} \tag{16}$$

This is the same form as (15) above.

The problem of matching coefficients is a difficult one. To reduce the complication somewhat, a few simplifying procedures can be performed. If R_1, R_2, and R_3 are all assigned nominal values, say 1, 5, 5, respectively, there are still enough variables to find a solution. (A low value for R_1 will help to ensure that a low gain value G can be used.) The three equations that result by equating coefficients are nonlinear and must

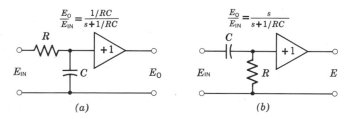

Figure 9.8 Simple active filters.

be handled by reiterative techniques. It is not necessary to start by assigning values only to R. All C values or combinations of R and C may be selected. No one combination improves the complexity of coefficient matching however. Once R and C values are determined an impedance transformation places R and C values into practical range.

9.6 SIMPLE FILTERS

Single RC or RL filter sections (simple real poles) can be found with a separate amplifier if the above technique is impractical. Figure 9.8 shows several simple circuits.

The individual filter sections described above can be designed to have specific damping and natural frequency values. In complex frequency terminology, complex pole pairs in the left-half s plane can be positioned by each filter section. Most standard filter types in use are composed of cascaded filter sections. A sixth-order Butterworth filter, for example, is formed by cascading three second-order filters, each with equal natural frequency but with staggered damping ratios. A fifth-order filter can be built by using a third-order section as in Figure 9.6 and a single second-order section as in Figure 9.2. The damping ratios or the pole locations are fully covered in the literature.

9.7 BUTTERWORTH, BESSEL, AND CHEBYSHEV

The standard low-pass filters are Bessel, Butterworth, and Chebyshev. The Butterworth filters have a flat frequency response without amplitude peaking, but they do exhibit 10 to 20% transient overshoot, depending on filter order. The Bessel filters have negligible transient overshoot but a sagging frequency response. From zero frequency to the —6-dB point, the amplitude-versus-frequency response is essentially unchanged as a function of filter order. (This assumes a normalized comparison where the — 3-dB points all occur at the same frequency.) Higher-order

Bessel filters do result in greater terminal slopes. Bessel filters are linear phase filters; that is, the time delay for sinusoids in the passband is essentially constant.

Chebyshev filters have both staggered natural frequency and damping. In the s plane, the poles lie on part of an ellipse. (The s-plane pole locations for a Butterworth filter are on part of a circle.) The Chebyshev filters have amplitude-versus-frequency ripple. The eccentricity of the ellipse and the peak-to-peak response ripple are related. Very small amounts of eccentricity with corresponding negligible amplitude ripple improve the "squareness" of the amplitude-versus-frequency response "knee." This improved response increases the transient overshoot however. The overshoot can be increased 50% by permitting less than 0.2% amplitude ripple. Any improvement in "squareness" reduces the phase linearity further, which causes the transient overshoot to increase.

Typical response curves and associated pole locations are shown in Figure 9.9 for each type of low-pass filter.[1]

9.8   NOTES ON ACTIVE FILTERS

Many applications require switching to modify the cutoff frequency or change the filter characteristic. It is more economical to switch resistors than capacitors. This has the effect of changing the impedance level. Doubling all resistors in an active filter reduces all frequency-dependent parameters by a factor of 2. If a large dynamic range of switching is needed, capacitors can be switched to cover decade ranges.

9.9   COMPONENT DISCUSSION

The equations that define filter performance show that element ratios are important. If a 10:1 resistor ratio is used and the switching requires another factor of 10, the total impedance span is 100:1. This may not appear too severe at first glance, but if parasitic phenomenon is to be less than a 1% effect, the range of permitted practical elements is limited. For this reason it is necessary to keep element ratios near unity if possible.

In filter amplifiers in which the input current and any dc offset are related resistors are limited to values that permit offset specifications to be met. Switching a voltage offset is usually avoided. If high values

---

[1] The pole locations for Butterworth filters provide a -3-dB point at $\omega = 2\pi f = 1.0$. The pole locations for other filter types are usually not normalized for -3-dB points.

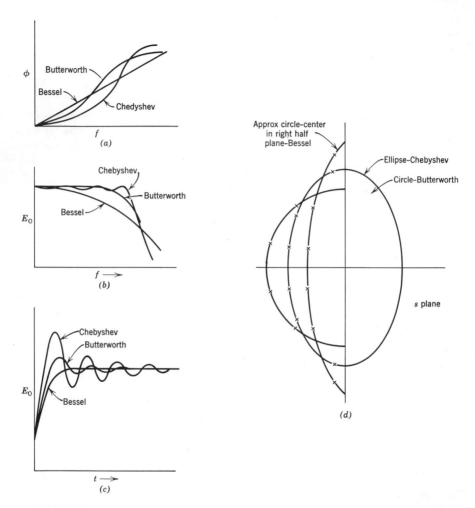

Figure 9.9 Characteristic filter data: (*a*) phase, (*b*) amplitude, (*c*) transient step response, (*d*) pole locations.

of resistance can be used, their shunt parasitic capacitances must be considered. If capacitance values range as low as 100 pF, a 1-pF parasitic capacitance across a resistor element limits the ratio of high frequency impedance to 100:1. This limitation can affect the response characteristics and should be considered.

The input impedance to an active filter is a function of frequency. Each section should have the current capability to accommodate the next segment. This load requirement is also related to the sequencing of filter segments. It is desirable to place the elements with large damping factors first to avoid excessive signal swing needs. Figure 9.9 shows that the damping ratio for a Bessel filter can be below 0.3. If this were the first filter section, the transient response peak of this segment alone would be over 30%. As a last section, no overshoot can occur. As an input segment, the output stage must provide 30% more current as well as 30% more voltage swing.

High-frequency low-gain filters with large signal swings require high slewing rates at the output. If potentiometric feedback is used, slewing limits may occur in the amplifier input stages. This limitation is not usually present in operational amplifier input stages.

Capacitors used as active filter elements should be measured for capacitance in the frequency range they are applied. The reader should review the material in Chapter 1 on dielectric absorption and on frequency-dependent capacitance to appreciate the problem. The designer should also recognize that the sensitivity to capacitance value can vary considerably. In low-damping cases, capacitance ratios and capacitance values can be very critical.

9.10   INTEGRATORS

A voltage function of time can be expressed in terms of a series of sinusoids. The integral of this waveform is the sum of integrals of sinusoids and is always of the form $\Sigma(A/\omega)$ cos $n\omega t$. The factor $1/\omega$ implies a response that is inversely proportional to frequency. This exactly states the nature of an integrator. In terms of an amplifier it is a device with a -6-dB/octave frequency-versus-amplitude slope at all frequencies.

Consider the amplifier in Figure 9.10. The input current is E_{IN}/R. This current flows in the capacitor C so that $E_0 j\omega C = E_{IN}/R$ or $|E_0/E_{IN}| = (1/RC)(1/\omega)$, which is exactly the requirement expressed above. Stated in integral form

$$E_0 = \frac{1}{RC} \int_0^T E_{IN}\, dt \qquad (17)$$

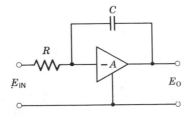

Figure 9.10   An integrating amplifier.

This circuit has several practical limitations. The assumption in the analysis is that $-A$ is very large. Obviously at some low frequency, the reactance of the capacitances will be large enough so that the feedback factor will fall below a useful level. This implies that the integrating time period T must be bounded. If $R = 1.0$ MΩ and $C = 0.1$ μF, then at 1 Hz the reactance of C is 1.6 MΩ. At 0.001 Hz, this reactance is 1600 MΩ. At this frequency, the amplifier has a voltage gain of 1600. If the open-loop gain is 160,000, the feedback factor is 40 dB. A single 0.001-Hz cycle represents an integrating period of 1000 sec. For an integrating period greater than 1000 sec, the loop gain is not adequate to maintain 1% integrating accuracy.

The start of an integration requires that the charge on C be zero. Charge can accumulate on C from the signal or from an internal source current. This source should contribute less than 1% error in 1000 sec. If 1% of full scale is 0.1 V, then $Q = CE = 10^{-8}$ pC and $I = 10^{-11}$ A for 1000 sec. Such a current level is difficult to realize on a guaranteed unattended basis. A zero control adjusted during a trial integrating period can be used to set the current to a near-zero value prior to a run.

An impedance level of 10^9 Ω requires that a quality capacitor be used, with care taken to ensure that the shunt resistance path is adequate. If the amplifier had infinite gain, the limiting integrating period would be determined by the integrating capacitor time constant. The integrating capacitor must also be low in dielectric absorption, or the charge released from absorption can also cause a large signal error. Dielectrics with low dielectric absorption are discussed in Chapter 1.

To facilitate the start of integration, a switch can be placed across C which opens upon command. If the integral can be evaluated, say, every 100 sec and a new integrating period started, the errors discussed above can be avoided. At the start of each integrating period, the capacitor is returned to zero charge.

Many integrating processes are unidirectional; that is, the integral is monotonically increasing. The scaling factor $1/RC$ must be such that the integral is within the available output voltage limits. If the capacitor is discharged when the output reaches a set value, a pulse can be generated to indicate that a block of integration has occurred. Pulse counting can be used to store the total integral value. The remaining voltage measures the fraction of a count. If the pulse rate is low or drops to zero, all of the problems discussed above must be considered.

Care must be taken to guarantee amplifier stability as the integrating capacitor forces the feedback factor to equal the open-loop gain at high frequencies. Because test signals are integrated, the observation of transient response (leading edge response) is obscured. The output slewing rate is bounded by the output current capability, and small signal levels are required before an observation can be meaningful. It is usually advisable to check the open-loop gain as a function of frequency to verify that the phase character has sufficient margin. Another check is to observe waveforms within the closed loop for square-wave inputs.

Some integrators are not required to operate over long time periods. Here an intentional shunt resistor can be placed across C. This provides a path for small internal source currents. The integral of signals without dc content is usable and the drift and error phenomenon discussed above is not relevant.

9.11   LOG AMPLIFIERS

The base-emitter junction voltage of a transistor obeys the following current-voltage relationship

$$V_{BE} = \frac{kT}{q} \log_e \frac{I_C}{I_S} \tag{18}$$

where I_C is collector current, I_S is the junction saturation current, k is Boltzmann's constant,[1] T is temperature in degrees Kelvin, and q is the charge on the electron. The saturation current I_S has been found to be nearly constant for matched adjacent transistors on a common substrate. The effect of I_S can thus be balanced out by using a differential transistor pair as shown in Figure 9.11.

The difference in base-emitter potentials is now

$$V_{BE_1} - V_{BE_2} = \frac{kT}{q} \log_e \frac{I_{C_1}}{I_S} - \frac{kT}{q} \log_e \frac{I_{C_2}}{I_S} = \frac{kT}{q} \log_e \frac{I_{C_1}}{I_{C_2}} \tag{19}$$

[1] Boltzmann's constant k is 8.6167×10^{-5} eV/°C.

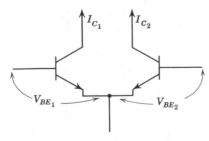

Figure 9.11   A balanced differential transistor pair.

The quality of semiconductor materials is such that this relationship is met over a 6 decade current range.

The relationship (19) is the basis for a log amplifier design. If both collector currents are known and one current is proportional to an input voltage, the base potential difference is proportional to the logarithm of the input voltage. An example of this technique is shown in Figure 9.12.

The first operational feedback amplifier ensures that $I_{IN} = E_1/R_1$ and this current flows in Q_1. The current in Q_2 is a constant reference value defined by Zener diode D_1 and transistor Q_3. Voltage E_L is the base difference of potential between Q_1 and Q_2 and is therefore

$$E_L = \frac{kT}{q} \log_e \frac{I_{C_1}}{I_{C_2}} = \frac{kT}{q} \log_e \frac{E_1}{R_1 I_{C_2}} = \frac{kT}{q} \log_e E_1 - \frac{kT}{q} \log_e \frac{1}{R_1 I_{C_2}} \quad (20)$$

Amplifier A_2 provides gain and offset to condition the signal E_L. If the offset voltage of A_2 RTI is $(kT/q) \log_e (1/R_1 I_{C_2})$ and the gain is q/kT, then the output voltage is exactly

$$E_O = \log_e E_1 = 2.303 \log_{10} E_1 \quad (21)$$

where E_1 is a positive voltage.

The use of a temperature-regulated substrate can keep T constant

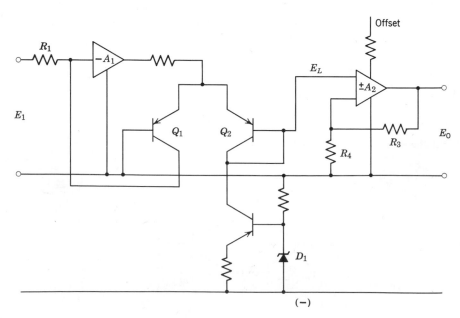

Figure 9.12 A log amplifier.

in (21). If the temperature is not regulated, the results vary proportional to absolute temperature. A 10°C change at room temperature causes a 3% change in the gain constant. A substrate that regulates the temperature to within 1° is usually quite acceptable in most applications.

9.12   LOG RATIOS

If I_{c_2} is made proportional to a voltage E_2, the voltage E_L is proportional to the log ratio of the signals E_1 and E_2. A current source proportional to E_2 is shown in Figure 9.13.

The voltage E_L is

$$E_L = \frac{kT}{q} \log_e \frac{I_{c_1}}{I_{c_2}} = \frac{kT}{q} \log_e \left| \frac{E_1 R_3}{E_2 R_1} \right|$$

provided that $I_{c_2} = -E_2/R_3$. If the gain of A_3 is q/kT and its offset is $kT/q \log_e (R_3/R_1)$, then

$$E_O = \log_e \frac{E_1}{E_2} = 2.303 \log_{10} \frac{E_1}{E_2} \tag{22}$$

Note that, as shown, E_1 must be a positive voltage and E_2 must be a negative voltage. If E_2 is positive, an inverting amplifier must be used to supply the correct polarity of current to Q_2.

9.13   MULTIPLIERS

The processes described in Section 9.11 can be extended to provide voltage or current multiplication. Consider the two matched transistor pairs in Figure 9.14.

For transistors Q_1 and Q_2,

$$E_L = \frac{kT}{q} \log \frac{I_1}{I_2}$$

and for transistors Q_3 and Q_4,

$$E_L = \frac{kT}{q} \log \frac{I_4}{I_3}$$

Equating like terms

$$\log \frac{I_1}{I_2} = \log \frac{I_4}{I_3} \tag{23}$$

This can also be written as $\log I_1 I_3 = \log I_2 I_4$ and thus $I_1 I_3 = I_2 I_4$.

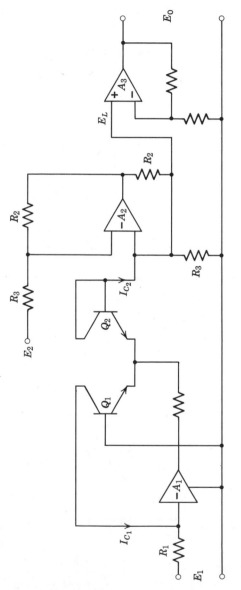

Figure 9.13 A circuit for taking the log ratio of voltages.

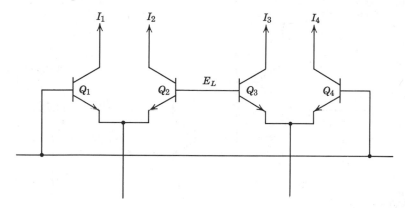

Figure 9.14 A multiplier configuration.

If I_1 and I_3 are supplied by operational feedback amplifiers proportional to voltages E_1 and E_3 and I_4 is sensed by an operational amplifier to produce E_0 (i.e., $E_0 = R_4 I_4$), then

$$\frac{E_1}{R_1}\frac{E_3}{R_3} = \frac{I_2 E_0}{R_4}$$

and

$$E_0 = \frac{R_4}{R_1 R_3 I_2} E_1 E_3 \tag{24}$$

If I_2 is supplied by an operational amplifier proportional to E_2, then

$$E_0 = \frac{R_2 R_4}{R_1 R_3} \frac{E_1 E_3}{E_2} \tag{25}$$

If $R_1 R_3 = R_2 R_4$, then $E_0 = E_1 E_3 / E_2$.

A full implementation of Equation (25) from Figure 9.14 is shown in Figure 9.15. Note that E_1 and E_2 must both be positive polarity voltages. It is also interesting to observe that (25) does not contain the factor kT/q which means that the multiplier is insensitive to temperature effects.

In practical circuits, diode clamps or limiting resistors such as R_0 ensure against component damage. Also, currents flowing in the multiplying transistors can unbalance the individual temperatures. If the transistor pairs are unbalanced, Equation (25) is no longer valid. This places a limit on the current levels that can be accommodated for a particular accuracy requirement.

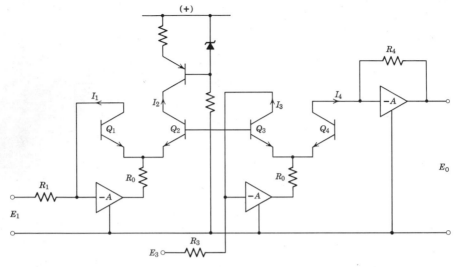

Figure 9.15 A multiplication circuit where $E_O = C E_1 \times E_3$.

The factors in (25) show that division between two variables is possible. When I_1 and I_2 are made proportional to voltages E_1 and E_2, then

$$E_O = \frac{E_1}{E_2} k \tag{26}$$

where $k = E_3 R_2 R_4 / R_1 R_3$. If $E_3 = R_1 R_3 / R_2 R_4$, then

$$E_O = \frac{E_1}{E_2} \tag{27}$$

Similarly, if $R_2 R_4 / R_1 R_3 E_2 = k_1$, then

$$E_O = k_1 E_1 E_3$$

and, if $E_2 = R_2 R_4 / R_1 R_3$, then

$$E_O = E_1 E_3 \tag{28}$$

an ideal multiplier.

9.14 SQUARE ROOT DEVICES

Several approximation techniques are available for generating the square root of a function. One approach that uses operational amplifiers requires a multiplier circuit in the feedback path. Such a circuit is shown

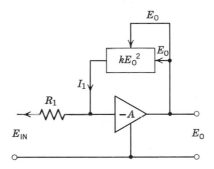

Figure 9.16 A square-root circuit.

schematically in Figure 9.16. The feedback balance requires $I_1 = I_2$. If the two input terminals to the multiplier are both connected to E_0, then $I_2 = kE_0^2$. Since $I_1 = E_{IN}/R_1 = I_2$, the combined equations yield

$$kE_0^2 = \frac{E_{IN}}{R_1}$$

or

$$E_0 = E_{IN}^{\frac{1}{2}}(kR_1)^{-\frac{1}{2}} \qquad (29)$$

PROBLEMS

1. Prove that a linear phase shift versus frequency is equivalent to a fixed time delay.
2. Design a low-pass filter with gain 1, $R_1 = R_2 = 10$ kΩ, $\zeta = 0.7$, and $\omega_n = 100$ rad/sec (see Figure 9.2).
3. Design a high-pass filter as in Problem 2.
4. An integrator is to have 1% frequency error at 0.1 Hz. The input R is 1 MΩ. What is the amplifier gain and the source current limitation? The integration gain $1/RC$ is to be 1000.
5. A low-pass second-order filter has a damping fraction of 0.2. The frequency response peaks near the natural frequency. What is the maximum gain if the dc gain is unity? Does the peak in amplitude response correspond to the 90° phase-shift frequency?
6. Derive an equation for damping fraction in a second-order filter as a function of peak-amplitude response.
7. Devise an equation for the gain of a low-pass filter at its 90° phase-shift point as a function of damping.

Index